Rolling stock
BR AND OWNER WAGONS

Colin J. Marsden

LONDON

IAN ALLAN LTD

Front cover: PAA No BIS 7987 owned by British Industrial Sand stands at King's Lynn during June 1983. *Barry J. Nicolle*

Below: Coupling arrangement between two air brake fitted freight vehicles, using the two pipe (main reservoir and brake pipe) air system. 1. Screw coupling, 2. Air brake pipe connection (painted red), 3. Main reservoir pipe connection (painted yellow), 4. Air brake pipe isolating cock (shown in open position), 5. Main reservoir pipe isolating cock (shown in open position), 6. Stowed screw coupling (coupling from one vehicle only used for coupling vehicles together), 7. Dummy lugs for air pipe when not in use.
British Railways

First published 1984

ISBN 0 7110 1403 5

Published by Ian Allan Ltd, Shepperton, Surrey; and printed by Ian Allan Printing Ltd at their works at Coombelands in Runnymede, England.

Introduction and TOPS

Rolling Stock Recognition 2: BR and Private Owner Wagons covers the present day wagon fleet (excluding departmental and service stock) by dividing the fleet into the TOPS letter groups, each group being given its own section — headed by a full list of types together with their appropriate three letter code. Each section is then illustrated with photographs of as many types and varieties that space has permitted. All vehicle types have not been illustrated as it is not the intention of this volume to cover every series of vehicle pictorially, but to provide the reader with a guide to the wagon types seen today. Number ranges of individual series have not been given as in some instances many different series apply to a type. In the near future the majority of unfitted vehicles (no continuous brake) will be withdrawn, together with a considerable number of vacuum braked examples, this occurring as modern purpose built air braked vehicles commence operation.

This book is based on the BR TOPS three letter wagon code system, so a brief description of the system is given. TOPS stands for *Total Operations Processing System* which is a computer based communications and information system. The TOPS computer for traffic control is a development of an American system introduced in the early 1960s for the Southern Pacific Transportation Company. In the closing years of the 1960s BR was actively involved in developing a computer based system for wagon and traffic control to ensure that (a) a constant supply of vehicles was available to BR and its customers (b) a trace could be made at any time to ascertain the whereabouts of a vehicle and the nature of its loading (c) maintenance cycles for stock and locomotives were adhered to. BR was faced with making a decision of what type of system to adopt, whether to develop a completely new method or to utilise experience gained by the US in their system. Eventually a modified development of the Southern Pacific computer was decided upon and given the same name as its American counterpart — TOPS.

The BR TOPS master computer situated in London, holds records of all BR locomotives, goods vehicles owned by BR or private companies and all passenger and NPCS vehicles; in other words all BR/Private railborne vehicles can be traced by any TOPS terminal. Each locomotive or vehicle has a permanent entry giving basic information such as size, weight, capacity and maintenance. This is joined by 'active' information about the present status of each vehicle, this being fed into the master computer from any one of a number of local offices with TOPS reporting facilities. To differentiate the vast number of wagon types in operation a three letter code system was developed: The first letter indicating the wagon or type group, this is often derived by the type of vehicle ie Group H is for hopper wagons, Group F for flat wagons. However there are a number of exceptions to this rule where more than one group is used for the same traffic purpose. The second letter of the code identifies the actual type or fleet within the Group, this may be by weight, size, duty or other distinguiching factor. The final letter indicates the brake type fitted and will be one of the following:

A	Air braked
B	Air braked with through vacuum pipe
F	Vacuum braked with Accelerate Freight Inshot (AFI)
G	Vacuum braked with AFI and through air pipe
H	Dual air and vacuum brake with vacuum AFI
O	Unfitted vehicle (no automatic brake)
P	Unfitted vehicle with through vacuum pipe
Q	Unfitted vehicle with through air pipe
R	Unfitted vehicle with through air and vacuum pipes
V	Vacuum braked
W	Vacuum braked, Air piped
X	Dual air and vacuum braked

The three letter TOPS code will be found on the wagon information panel or painted on the vehicle's side, usually in white. When trains are being formed it is most important for this code to be fully understood as trains of mixed types may not be compatible together.

When a TOPS office, area freight centre or motive power depot that is fitted with a terminal requires information about an individual locomotive/vehicle the operator keys in a code for the information required, followed by the locomotive or wagon number including any prefix letters. The required information will then be given either on a video display unit (VDU) or by a hard printer. The TOPS computer is also used for transmitting information of complete train loads from yard to yard, enabling the receiving point to be prepared to accept and deal with the train and its load. Whenever a wagon is moved between yards, whether loaded or empty, all information is given to the master computer. Every wagon to arrive in a yard is checked and its number again given to the computer with details of its next duty, if known, enabling the master computer to schedule the vehicle for its next movement. Maintenance cycles can then be programmed from the TOPS information. Before each train commences a journey the following information is placed on TOPS enabling a train preparation form to be produced by a computer hard printer before departure, and a copy given to the train crew.

a Wagon numbers (including prefixes) in the same order as they occupy behind the locomotive
b The loading of all vehicles
c The individual wagon type (by TOPS code)
d The length of the train (excluding motive power), progressively totalled from the front wagon by wagon
e The weight of the train (including motive power) progressively totalled from the front wagon by wagon
f The brake force (in tonnes)
g The destination of each wagon (by TOPS figure code)
h Route directive code

Below: It is normal for all wagons to carry a maker's plate. Those fitted to older vehicles give the actual wagon number, weight, location of build and year of manufacture, whereas alloy plates fixed to modern vehicles give little information. This Shildon 1979 plate was photographed on the side of a British owned privately operated ferry wagon. *Colin J. Marsden*

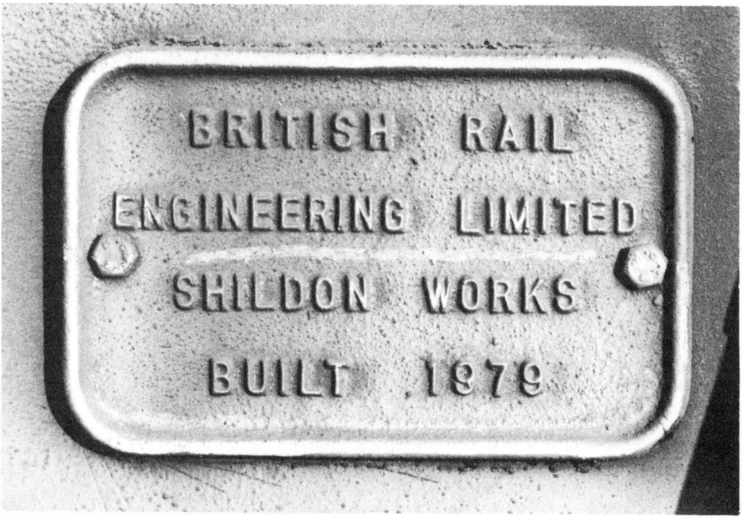

Above: Various styles of wagon labels have been applied over the years, all giving the same information. This plate is from a 45 ton hopper wagon. In the left column the total weight in tons is shown, with underneath, the Brake Force VB, Route Availability RA, and Maximum Speed. To the right are four columns headed by the letters H (Heavy loaded), M (Medium loaded), L (Light loaded) and E (Empty); from this table the guard or train preparer can calculate the brake force available to the driver, the route the train can operate over, and the maximum permissible speed of the train. *Colin J. Marsden*

After each train is despatched the TOPS computer is informed and it then gives a *consist* message to all stopping points for that particular service, giving yard staff a full breakdown of the train due, including motive power, and any repairs or attention needed to vehicles in the formation. Empty wagons are also monitored by TOPS enabling the maximum utilisation of vehicles and keeping empty mileage to an absolute minimum.

In addition to BR depots, TOPS offices and Area Freight Centres having access to TOPS computer information, a number of larger private operators have terminals where information can be obtained about their own vehicles and cargos, however full BR wagon/locomotive information is not available from these terminals.

The TOPS computer is a large system with vast information recallable in a number of ways. It can produce data about one wagon, a single locomotive's maintenance history, a full list of vehicles at a particular location, or find vehicles fitted with a special modification (ie locomotives fitted with snowploughs), to name but a few. All this information is governed by the code numbers/letters fed into the machine by the operator. However like any computer based system TOPS can only be as accurate as the information fed to it, and therefore cannot be fully relied upon for 100% accuracy at all times.

I would like to record my thanks to the many railway staff who have assisted with this project, Procor (UK) Ltd who provided assistance with a number of private owner vehicles, and the many railway photographers who kindly turned their cameras to wagons in answer to my pleas, to Mrs J. W. Marsden for typing the manuscript work, Alan Butcher for the line drawings and lastly to my publishers for giving me complete freedom in the preparation of this volume.

Colin J. Marsden
Surbiton
February 1984

Wagon Signs and Markings

Wagon numbers are usually applied in white or black paint on BR-built pre-air braked stock prefixed by the letter B and usually followed by six numbers. Vehicles built prior to nationalisation are prefixed by the company letters. In addition to the number the capacity, tare and wheel base are all to be found on the side.

A new wagon numbering system was introduced during the early 1970s without a prefix letter for modern air braked wagon stock. Private owner vehicles are numbered in company series prefixed by the owner's initials. Some of the most common symbols seen on wagon stock are shown:

XP — Denoting that the vehicle is permitted to operate in a parcels or passenger train formation, and therefore must have a wheel base of more than 10ft 0in. This symbol was normally applied in white surrounded in a box.

/ — A white diagonal at the end of a mineral wagon at 45° indicates that the vehicle has an end door at that end.

V — Mineral wagons having doors in the floor, have a white painted V in the centre of the vehicle's side.

ϟ — The overhead wire electric symbol in red on a white background; usually only applied to vehicles that convey ladders enabling staff to climb above wagon height.

⚓ — The anchor symbol indicates that the vehicle can operate through to the continent via train ferries.

★★ — The two star symbol found on tank wagons indicates the maximum speed is 60mph.

Alterations to Wagon Codes

As this book went to press several alterations to the TOPS wagon code system were announced by BR to permit the expansion of the private owner wagon and tank codes — as an ever increasing number of PO wagons are placed in traffic. The main alterations are:
Group J to be amalgamated with Group B
Group K to be amalgamated with Group S
Group X (Part) to be amalgamated with Group F
Group X (Part) to be amalgamated with Group B
Group U to be amalgamated with Group O

A train formed of various steel coil carrying vehicles of groups J and K, including some with nylon hoods, passes through Peartree station headed by Class 31 No 31.279 on 1 July 1983. Trains of this type are usually restricted in their length to a maximum of between 15-20 vehicles on weight grounds, as some of the vehicles could weigh up to 70 tonne when loaded.
Colin J. Marsden

Group B: Bogie Steel Carrying (Excluding Coil)

Code	Description	Capacity	Brake Type	Remarks
BAA	Bogie steel 40ft	77.5-78.5t	Air	
BAB	Bogie steel 40ft	77.5t	Air-Vacuum piped	
BBA	Bogie steel 50ft	73.5-75.5t	Air	
BCO	Bogie bolster C	30.5t	Unfitted	
BCV	Bogie bolster C	30.5t	Vacuum	
BDA	Bogie bolster D	42.5-58t	Air	
BDO	Bogie bolster D	42.5t	Unfitted	
BDV	Bogie bolster D	42.5t	Vacuum	
BEV	Bogie bolster E	32.5t	Vacuum	
BHV	Bogie bolster H	30.5t	Vacuum	
BKA	Bogie steel—Kinky beam	75.5t	Air	
BKB	Bogie steel—Kinky beam	75.5t	Air, Vacuum piped	
BPA	Boplate E	58.5t	Air	
BPO	Boplate E	42.5t	Unfitted	
BPV	Boplate E	42.5t	Vacuum	
BQV	Bogie Bolster Q	30.5t	Vacuum	
BRA	Borail	51t	Air	
BRP	Borail	51t	Vacuum piped	
BRV	Borail	51t	Vacuum	Some vehicles have bolsters while others have curb rails
BTV	Bogie bolster T	30.5t	Vacuum	

Below: BAA, BAB. A number of bogie steel-carrying wagons were constructed during the 1970s, the majority capable of carrying loads in excess of 50 tonne. 305 vehicles were built from 1972 capable of carrying 77.5 tonne and are classified BAA or BAB. Vehicles carrying the code BAA Nos 900000-900124, 900200-900305 are fitted for air brake operation, whereas vehicles Nos 900125-900198 are fitted for air brake operation and carry a through vacuum pipe, therefore classified as BAB, A BAA vehicle is shown at the rear of the train in this illustration, with BRA and BBA vehicles in front. BAB not illustrated. *Colin J. Marsden*

BBA, BCO, BCV

Above: BBA. All modern purpose built steel carrying stock is fitted with corrugated decking which aids the effective dissipation of heat from hot steel loaded on to wagons in the foundries. The major difference between the BBA vehicles and the BAA/BAB type is their length, 50ft compared to 40ft. Nearly 700 vehicles of this type are in service in the number range 910000-910691, and can normally be found operating in block train formations mainly in the North of England. *Colin J. Marsden*

Below: BCO, BCV. During the 1950s several hundred bogie bolster vehicles were constructed to different sizes for various applications, some were fitted with vacuum brake equipment, while others were unfitted. The BCO or bogie bolster C, not fitted with a continuous brake, is now almost extinct. These vehicles can carry 30.5 tonne of merchandise evenly spaced. BCO No B943616 illustrated here, is being used as a spacer vehicle between two concrete beam wagons. BCV not illustrated. *British Railways*

Bottom left: BDA. A substantial fleet of bogie bolster D (BDA) vehicles fitted with air brakes were introduced from 1975 and to date over 1,500 have been built, however a number are now used by the engineering departments and classified as departmental stock. These bogie steel carrying vehicles can convey between 42.5 and 58 tonne, depending on their build. The bodies are fitted with up to 10 removable stanchions and tethering points which can be used according to the load carried. BDA No 951097 shown here, is painted in Railfreight red livery; the white wheels on the bogies are for applying and releasing the handbrake. *Michael J. Collins*

Above: BDV. Two different bogie designs can be found under the BDV fleet; early built vehicles have plate frame bogies whereas later built examples have Gloucester design bogies. The plate bogie is displayed here on BDV No B927705. The cross bolster members of these wagons are in a fixed position, but on the BCO/BCV fleet bolster members are moveable. *Colin J. Marsden*

Below: BEV. In 1961-62 bogie bolster E vehicles were constructed in two lots, these vehicles can carry 32 tonne mounted on four movable cross bolsters. All BEV vehicles are mounted on Gloucester style bogies. These vehicles are 32ft 0in over headstocks and are used to convey loads of up to 30ft 0in in length, when in tare condition they weigh 16 tonne. BEV No B923564 is shown here. These wagons were the final bogie bolster development prior to the introduction of the air braked fleets. 1,000 of these vehicles have now been rebuilt for departmental use and named Turbot. *Barry J. Nicolle*

BKA, BKB, BPA, BPO, BPV

Top: BKA, BKB. The fleets of BKA and BKB bogie steel wagons fitted with a kinky beam, were converted from members of the BAA/BAB fleet and included the fitting of longitudinal beams for carrying steel coils. Vehicles can carry up to 75.5 tonne and can usually be found operating in block train formations. No 900111 is illustrated here at Cambridge.
Barry J. Nicolle

Above: BPA, BPO, BPV. A fleet of 80 BPA steel plate carrying vehicles was introduced from 1980, these are the modern version of the vacuum braked Boplate E, some of which are still in traffic. The BPA is designed for the conveyance of steel plates and other similar steel traffic. Each wagon is fitted with cross floor battens and full length drop sides to facilitate easy loading and unloading. When fully loaded the BPA type can carry 58.5 tonne and weigh 21.4 tonne when in tare condition. BPA No 965014 is seen here at Reading.

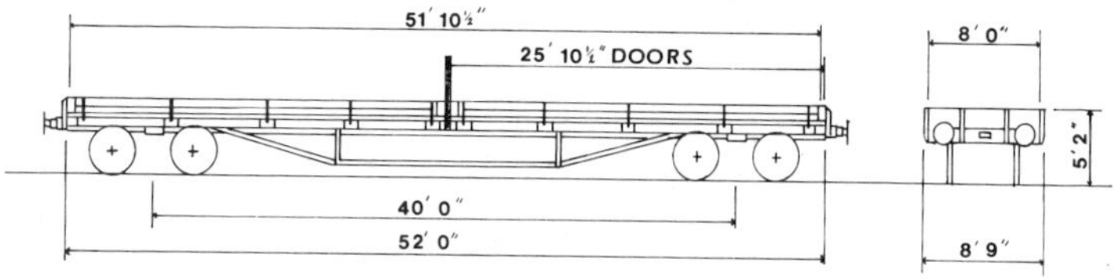

BPA
Tare: 21.4 tonne
Carrying capacity: 58.5 tonne
Minimum curve: 30m
Body: Fixed steel ends, wooden drop sides
Floor: Wood

Above: BRA. 150 Borail wagons were constructed during 1981 and classified as BRA. Each vehicle is fitted with six cross bolsters with movable vertical support stanchions, and intended to convey steel sections or rails. The majority of these vehicles have now been transferred to the departmental fleet for permanent way use and reclassified as YLA. BRA No 967586 is shown here loaded with 60ft rails at Woking. *Colin J. Marsden*

Below: BRV. The Borail fleet or BR group under the TOPS system, are intended for the conveyance of extra long loads, ie rails, concrete structures or engineering plant. Several batches of this type, fitted with vacuum brakes, have been built at various workshops all to a similar design but with minor detail differences. One striking feature of the type is the size of the bogies, the largest fitted to any freight stock in terms of wheelbase. No B946219 is shown here, built by BR at Derby in 1961. *Colin J. Marsden*

Group CA: Brake Vans (Traffic Department)

Code	*Description*	*Brake Type*	*Remarks*
CAO	Traffic Brake Van	Unfitted	
CAP	Traffic Brake Van	Vacuum Piped	
CAR	Traffic Brake Van	Unfitted — dual piped	

Group C under the TOPS classification system is for Covered Bulk Carriers, however at the beginning of this group using the letters CA, are Traffic Brake Vans. Only three types exist CAO, CAP and CAR, however there are a number of different designs. It should also be noted that a number of Traffic Brake Vans have been transferred to the Departmental fleet without any alteration except to the TOPS coding.

Below: CAO. The BR standard brake van is of the 20 tonne type and owes its design to the LNER, who first built 16ft wheelbase examples for fast freight services during the early 1940s. All LNER built vehicles were fitted with vacuum brake equipment, however when BR carried on the design, albeit with minor detail alterations during the 1950s, a number of unfitted examples were constructed and these now conform to TOPs code CAO. No B953438 is shown here which was one of a batch of 260 built under lot No 3012 in 1957. *Colin J. Marsden*

Above: CAO. The brake van on trains is supplied to convey the guard and is fitted with observation ports protruding from the side of the vehicle, and end viewing verandahs, these having a thick concrete floor to add weight to the vehicle. There is a coal stove to provide heat for the guard during the winter months and the chimney can be seen on the roof. Between 1948 and 1962 over 5,000 brake vans of various designs were introduced. No B953045 of type CAO is shown here at Derby. *John Tuffs*

Below: CAP. This 20 tonne standard brake van is of the piped type, meaning that it carries vacuum pipes but is not itself brake fitted. On these vehicles the heads and solid part of the vacuum pipe should be painted white, but after only a few days in traffic they soon become discoloured. A number of variations exist within the standard brake van fleet in terms of grab rails, lamp iron positions; and a number of later examples have hydraulic buffers and roller bearing axles. In this illustration No B950656 has only standard buffers and oil filled axle boxes. *Barry J. Nicolle*

Above: CAP. The other common type of brake van seen in traffic today is of LMS design, where the end verandah extends the full length of the frame unlike the BR standard design that has protruding ends. Large numbers of this design were constructed prior to nationalisation and further examples were built after 1948 both of the fitted and unfitted types. The box section between the wheels carries steel scrap to give additional weight to the otherwise light weight vehicle. Piped No B950015 is seen here at Hartlepool during 1982 and is one of a batch built at Derby during 1950. *Colin J. Marsden*

Below: CAR. A number of BR standard type vehicles have in recent years been fitted with both vacuum and air piped equipment and classified as CAR. Although the brake van itself has no automatic brake equipment, the attached trains' vacuum and air brakes could be applied by taps in the van. These vans often operate in conjunction with merry-go-round coal trains, and can be recognised from vacuum braked or piped examples by additional air pipes on the buffer beam, yellow painted panels on the ends and side and the slogan 'air piped'. No B954625 is seen here at Derby. *John Tuffs*

Above: This view of the stabling point at Hartlepool shows a Class 37 locomotive with four brake vans, three BR standard design vehicles and one of former LMS design. The need for guard's brake vans on freight trains is now rapidly disappearing with the implementation of fully fitted trains. The purpose of a brake van is to convey the guard at the rear of unfitted trains to ensure that the train is travelling in a correct manner and to provide guard's accommodation on fitted trains when he is required to be at the rear of the train for shunting or other duties. *Colin J. Marsden*

Below: ZSQ. Although this volume of *Rolling Stock Recognition* does not deal with departmental stock this ZSQ has been included to show that special applications can be given to traffic stock. Here standard 20 tonne brake van B951496 has been transferred to the departmental fleet and converted to a match wagon for modern high-density EMU stock of Classes 317/455 and fitted with low level tightlock couplings at one end. The vehicle is branded to work between Cricklewood-Bedford and Wolverton only, but was photographed at Waterloo with the first Class 455 EMU to arrive on the SR late in 1982. The vehicle carries a Q brake classification indicating that it is air piped only.
Colin J. Marsden

Group CB-CZ: Covered Bulk Carriers

Code	*Description*	*Capacity*	*Brake Type*
CBA	Covered hopper	31t	Air
CCO	Covered sand hopper	24-25t	Unfitted
CCP	Covered sand hopper	24-25t	Vacuum piped
CCV	Covered sand hopper	24-25t	Vacuum
CGO	Covered grain	20.5t	Unfitted
CGP	Covered grain	20.5t	Vacuum piped
CGV	Covered grain	20.5t	Vacuum
CHO	Covered hopper	24.5t	Unfitted
CHP	Covered hopper	24.5t	Vacuum piped
CHV	Covered hopper	24.5t	Vacuum
CPV	Presflo hopper	17.5-22.5t	Vacuum
CPW	Presflo hopper	22.5t	Unfitted vacuum & air piped
CQV	Prestwin	20.5t	Vacuum
CSA	Ash hopper	21.5t	Air
CSV	Ash hopper	21.5t	Vacuum
CXV	Gun powder wagon	7-11t	Vacuum
CZO	Alumina	17.5t	Unfitted
CZP	Sugar	23.5t	Vacuum piped
CZV	Sugar	23.5t	Vacuum

Below: CCO, CCP, CCV. A large number of 24 tonne bulk powder or *Covhop* wagons were built during the 1950s and early 1960s for the conveyance of powdered chemicals. Many are now utilised for soda-ash traffic but the one illustrated here is for British Industrial Sand and destined for use in the manufacture of Rockware glass. A number of unfitted, vacuum braked and piped examples were built mainly by BR at Derby. The vehicle illustrated is of the piped type and fitted with roller bearing axles. *Colin J. Marsden*

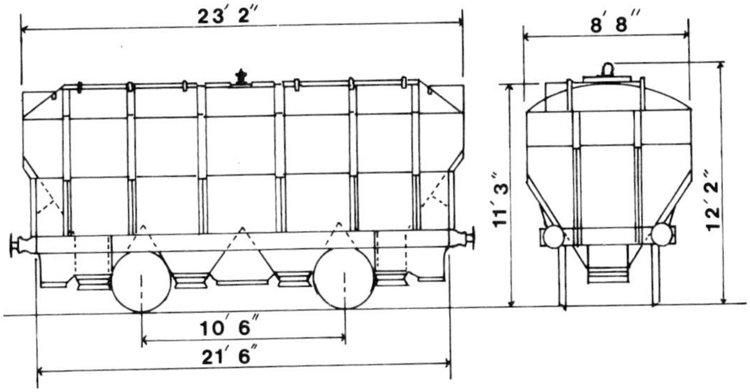

CCO, CCP, CCV
Tare: 11 tonne
Carrying capacity: 24.5 tonne
Minimum curve: 21m
Body: Steel

Below: CGO, CGP, CGV. The present fleet of covered grain hoppers are classified as CGO, CGP or CGV (depending on the brake fitting) and were built during the mid 1950s as a development of the LMS and GWR types. The grain is loaded through two roof ports, and discharged via floor doors, controlled by a wheel (on other side of vehicle). The majority of vehicles are fitted with roller bearing wheelsets. CGO No B885820 stands at Newmarket in 1981 and is one of a fleet built by Pressed Steel in 1955. *Michael J. Collins*

Above: CPV, CPW. These wagons are often referred to as *Presflo* vehicles, due to air pressure being needed for the discharge of the powdered commodity being carried. The majority of CPV or CPW types are allocated to cement traffic, although other powdered goods can also be transported. Between 1954 and 1962 over 1,700 of these vehicles were built. CPV No B887952 shown here, was built in 1958 by Gloucester Carriage & Wagon Co. CPW type not illustrated. *Colin J. Marsden*

Below: CSA, CSV. During the mid-1960s a fleet of 21 tonne ash Presflo wagons were introduced, mainly for the conveyance of flyash from power stations to industrial sites. Both air and vacuum braked examples were constructed and could normally be observed operating in block train formations. CSA No B873989 built in 1965 is photographed here at Nuneaton. All are fitted with roller bearing wheel sets and can be recognised from the standard Presflo wagon by their slightly larger size. CSV type not illustrated. *Colin J. Marsden*

Above: CXV. During the 1950s, 135 gunpowder wagons were constructed by BR at the former GWR works at Swindon, closely following the former GWR design of vehicle. Fleets of both fitted and unfitted vehicles were constructed but today only fitted examples remain. Two types — the 7 and 11 tonne varieties, are now in traffic the latter type of which is shown here in the form of No B887152 built in 1959. *Paul Shannon*

Below: CZP, CZV. Part of the *Covhop* fleet under TOPS type CC are set aside for the conveyance of sugar traffic and classified as CZ. Two brake types are in use, vacuum piped (CZP) and vacuum fitted (CZV). The best way to recognise this CZ group from the CC type is by reference to the TOPS code on the body size. CZV No B870840 is seen here at March during 1981. CZP type not illustrated. *Michael J. Collins*

Group F: Flat Wagons

Code	*Description*	*Capacity*	*Brake Type*	*Remarks*
FBB	Conflat AB	25.5t	Air, Vacuum piped	
FEV	Conflat E	42.5t	Vacuum	
FFA	Freightliner Inner	52-62t	Air	
FGA	Freightliner Outer	52-62t	Air	
FGB	Freightliner Outer	52-62t	Air, vacuum piped	
FHA	Lowliner	49-63t	Air	
FIX	Carfit C (RIV standard)	20.5t	Dual	
FJA	Skeletal 60ft	52t	Air	Ex Freightliner
FJB	Skeletal 60ft	52t	Air, vacuum piped	
FLV	Conflat L	14t	Vacuum	
FMA	Freight Flat	42t	Air	Timber Floor
FPB	Conflat	22.5-31.5t	Air, vacuum piped	
FSV	Conflat ISO	2.5t	Vacuum	
FVV	Carflat	10t	Vacuum	Including motorcar flat
FVW	Carflat	10t	Vacuum, air piped	Including motorcar flat
FVX	Carflat	10t	Dual	Including motorcar flat
FWV	Boflat	42.5t	Vacuum	
FZA	Lowliner	10t	Air	

Below: Freightliner train in action. The flat vehicles of the FF group forming the freightliner fleet operate on specific duties, normally in block formations made up of five vehicle lengths. Containers of various shapes and sizes will be formed on one train, belonging to a number of operators. A northbound freightliner train passes Wetmore, Burton on Trent, with only five containers on board, headed by Class 45, No 45.060. *Colin J. Marsden*

Above and right: FFA, FGA. Probably the greatest advance in freight traffic in this country came in 1963 when the first prototype vehicles for *Freightliner* services were introduced, with the production vehicles following in 1965. Freightliner sets are usually formed in block formations of five vehicles, semi-permanently coupled between inner vehicles by bar couplings, with conventional drawgear at the outer ends of the end vehicles. There are two types of Freightliner vehicle in existence, the FFA (Freightliner outer vehicle) and the FGA (Freightliner inner vehicle). The only difference between the two is the omission of standard drawgear from inner vehicles. All are equipped with air disc brake equipment and mounted on Gloucester style bogies. Locating pins for standard length Freightliner containers are fixed to the vehicle frame which twist and lock into the container when lowered on to the flat. The upper illustration shows a FFA vehicle parted from the train, whilst the view on the right shows an end type FGA vehicle from the draw gear end. Between 1963 and 1972 over 2,100 freightliner vehicles were constructed. *Colin J. Marsden*

Freightliner Container Types

A number of different freightline container types are used on freightliner services. The table below gives a brief description of the majority of types.

External Length	*Type*	*Special Features*	*Tare*		*cu ft*	*Payload*		*Internal Length*	*Internal Width*	*Internal Height*
			T	*C*		*T*	*C*			
Box 10ft	G	Roller shutter one side	1	5	505	8	15	9ft 2$\frac{1}{2}$in	7ft 7$\frac{1}{4}$in	7ft 2$\frac{1}{2}$in
Box 20ft	L		2	2	1,085	17	17	19ft 4$\frac{1}{2}$in	7ft 7$\frac{1}{4}$in	7ft 2$\frac{1}{2}$in
	L	8ft 6in high	2	3	1,162	21	5	19ft 3$\frac{3}{4}$in	7ft 7$\frac{3}{4}$in	7ft 10in
Box 30ft	X	Top and end door hatches	3	5	1,618	21	15	29ft 4$\frac{3}{4}$in	7ft 8in	7ft 1$\frac{3}{4}$in
Box 30ft	N		3	1	1,618	22	2	29ft 5$\frac{1}{2}$in	7ft 7$\frac{1}{4}$in	7ft 2$\frac{7}{8}$in
	N	8ft 6in high	3	4	1,757	21	17	29ft 4$\frac{3}{4}$in	7ft 7$\frac{3}{4}$in	7ft 9$\frac{3}{4}$in
Box 40ft	F	8ft 6in high	4	3	2,366	26	1	39ft 5$\frac{1}{4}$in	7ft 7$\frac{3}{4}$in	7ft 10in
Curtain -sided 30ft	BOALLOY	Load bearing curtains	2	16	1,643	20	0	29ft 5in	7ft 10in	7ft 1$\frac{1}{2}$in
Open 20ft	W		1	14	—	18	6	19ft 5$\frac{1}{4}$in	8ft 2in	4ft 2in
Open 20ft	H		1	10	—	18	10	19ft 2$\frac{1}{2}$in	7ft 7$\frac{1}{4}$in	3ft 0in
Open 20ft	T	Coil Carrier	2	14	—	23	5	18ft 9in	8ft 1in	3ft 0in
Open 30ft	Y		2	8	—	22	12	29ft 6in	8ft 2in	3ft 2in
Open 30ft	K		3	1	—	21	19	29ft 3$\frac{1}{4}$in	7ft 7$\frac{1}{4}$in	3ft 0in
Open 30ft	C	High loading potential	2	17	—	25	3	29ft 6$\frac{1}{2}$in	8ft 1$\frac{1}{2}$in	3ft 1in

Note: Unless specially noted, all the above covered containers are 8ft 0in high and 8ft 0in wide externally. All have end doors.

Above: FIX. A fleet of Carfit C wagons were constructed during the late 1950s conforming to RIV standards for the conveyance of vehicles between England and the Continent. Vehicles can take a 20 tonne load and are fitted with dual brake equipment. No B748112 also carrying European number 2170 414 0 002-9 is photographed loaded with a tank at Cambridge during early 1981. *Paul Shannon*

Below: FMA. Some early built freightliner vehicles were rebuilt during 1976 to FMA wagons and were primarily converted for the transport of lorry chassis between Bathgate and Birmingham. They are formed in three vehicle blocks, two outer and one inner vehicle and semi-permanently coupled by a bar coupling. At the outer ends of the outer vehicles there is a ramp to ease offloading; floors are of wooden cross members. *British Railways*

Above: FVV, FVX. The FV group consists of Carflat wagons, the majority operating under the *Motorail* tradename. These vehicles have coach style underframes with wooden decking to accommodate two or three motor cars; hinged panel ends drop down to cover the buffing and draw gear to enable cars to be end loaded and driven the length of the coupled train. Tethering of cars is normally by wheel clasps or nylon leash. Vehicles have buck-eye type couplings and are provided with a standard pullman rubbing plate above. A variety of passenger coach pattern bogies will be found under these vehicles. FVV No B745031 is shown here. FVX are of similar design but not illustrated.
Colin J. Marsden

Below: FZA. Five lowliner vehicles classified under TOPS as FZA were introduced during 1970 for the conveyance of commercial vehicle chassis. Of the five, two were end units having buffing gear and loading ramps, and the other three were intermediate cars with internal couplings of the bar type. All vehicles had removable side panels. It will be noted from the illustration that when the screw coupling is not in use it is stowed on a bracket under the buffer, and not on the coupling shank. When built these vehicles were intended to operate with normal freightliner services. *D. Simmonds*

Group H: Hoppers

Code	*Description*	*Capacity*	*Brake Type*	*Remarks*
HAA	MGR hopper	32.5t	Air	45mph when empty
HBA	Mineral hopper	32.5-38t	Air	45mph when empty
HCO	Coke hopper	20.5t	Unfitted	
HCP	Coke hopper	20.5t	Vacuum piped	
HDA	Coal hopper	32.5t	Air	60mph when empty
HEA	Mineral hopper	32.5t	Air	60mph when empty
HJO	Ironstone hopper	26t	Unfitted	
HKV	Ironstone hopper	33.5t	Vacuum	
HSO	Coal hopper S	13t	Unfitted	
HTO	Coal hopper XX	21.5-25t	Unfitted	
HTP	Coal hopper XX	21.5t	Vacuum piped	
HTV	Coal hopper XX	21.5-25t	Vacuum	
HUO	Coal hopper XXV	24.5-25t	Unfitted	

Below: HAA. Many thousands of HA type hoppers are in operation working on merry-go-round (MGR) coal duties formed in trains of 30-36 hoppers. The purpose of the MGR system is to run the service on a practically non-stop basis, slowly moving through the loader at the colliery, transporting its load to the power station discharge facility and slowly passing over the unloading equipment, where the bottom doors on the hopper automatically open/close when passing over the discharge equipment. HAA No 353639 is shown here, the automatic discharge equipment being visible on the frame above the nearest wheel set. *Colin J. Marsden*

Above: HAA. Prior to the introduction of the TOPS wagon code system the HAA fleet were classified as *HOP AB* vehicles which are fitted with the two pipe air system and able to operate at up to 45mph. *HOP AB* No B350397 is illustrated in original condition.
British Railways

Note: The majority of HAA vehicles are being rebuilt with modified underframe equipment, enabling them to operate at 60mph and reclassified as HDA.

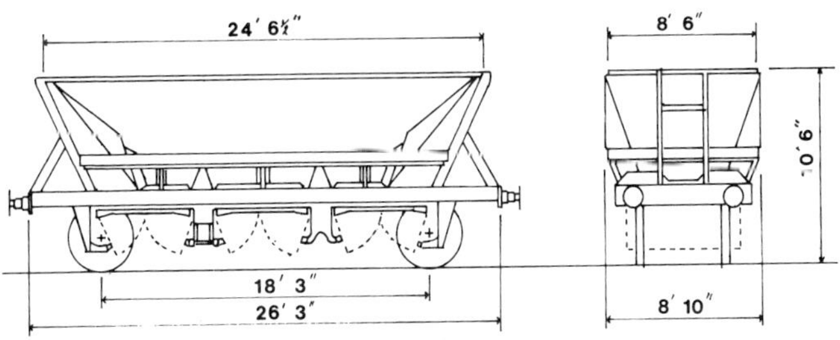

HAA, HDA
Tare: 14.5 tonne
Carrying capacity: 32.5 tonne
Cubic capacity: 32.7cu m
Minimum curve: 41m
Body: Steel

Above: HBA. Non-automatic discharge air braked hoppers are classified as HBA, often operating in block formations or on wagon load services. The capacity is 32.5 tonne; air disc brakes are fitted and two manually operated underslung hopper doors. No 360315 is illustrated here. *Colin J. Marsden*

Below: HBA. Unlike the HAA fleet HB vehicles have end ladders for vehicle inspection. Bauxite livery is usually applied to these vehicles, however a number have recently emerged in *Railfreight* red/grey livery with the *Railfreight* logo on the side. The maximum speed for this type is 45mph. No 360975 is illustrated here. *Colin J. Marsden*

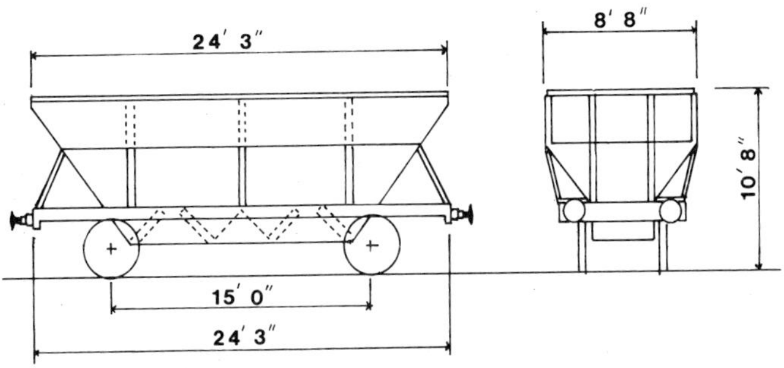

HBA, HEA
Tare: 13.3 tonne
Carrying capacity: 32.5 tonne
Cubic capacity: 31cu m
Minimum curve: 31m
Body: Steel

Note: The majority of HBA vehicles are being rebuilt with modified underframe equipment enabling them to operate at 60mph, and reclassified as HEA.

Below: HDA. A number of converted HAA and recently constructed air braked hopper wagons are classified as HDA, these are the same as the HAA type except that maximum speed is 60mph. All bodies of both HAA and HDA type wagons are in unpainted steel with black underframes. No 368198 is illustrated here. *Colin J. Marsden*

Above: HEA. To permit the wagon load to travel at 60mph a number of Railfreight air braked HBA hoppers have been modified in recent years and reclassified as HEA; shown painted in red/grey livery with a white logo and Railfreight sign is No 361873.
Barry J. Nicolle

Below: HEA. To improve the riding quality vehicle No 361798 was modified by the Research & Development section at Derby with revised suspension hangers. The vehicle was photographed at Derby during 1982. *Colin J. Marsden*

Above: HJO, HJV. This type can be described as the BR standard general purpose hopper capable of carrying 24.5 tonne of merchandise; well over 2,000 of these were constructed between 1956 and 1959. Today the numbers have greatly diminished and those that do survive are destined for specific usage, ie: the vehicle illustrated HJV No B439743 is branded for limestone traffic. The wheel above frame height to the left of the vehicle, is to control the floor mounted hopper doors. It is interesting to note that two vacuum brake cylinders (mounted at the near end of the vehicle) are provided. The HJO type is identical except for brake gear, not illustrated. *Colin J. Marsden*

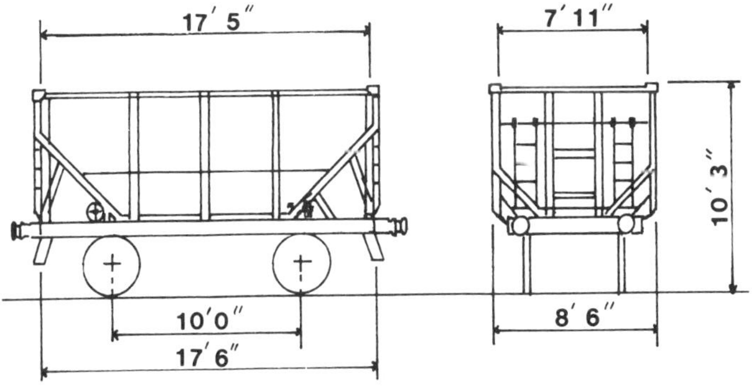

HJV
Tare: 11 tonne
Carrying capacity: 26 tonne
Cubic capacity: 18.77cu m
Minimum curve: 21m
Body: Steel
Floor: Steel

Above: HJV. Of slightly different design to the previous type is this HJV hopper designed for the conveyance of Ironstone. These can usually be found loaded with sand and often vehicles are branded for sand traffic only. This view clearly shows the vacuum brake cylinder and end ladder arrangement. This vehicle No B437746 is incorrectly carrying a TOPS code of HKV (reserved for 33.5 tonne hoppers) when photographed at Norwood.
Colin J. Marsden

Below: HTO. This design of standard coal hopper capable of carrying 21 tonne of coal is a development of an LNER design, who were the leaders in bottom discharge hopper wagons. A number of various designs and detail differences exist within this large fleet. The vehicle illustrated is of the unfitted type and therefore classified as HTO. No E307787 has a tall handbrake lever typical of LNER style, but has been rebodied in recent years.
Colin J. Marsden

Above: HTV. Since nationalisation more than 23,000 21-tonne standard hopper wagons have been built, although all would not have been in service at the same time. This view of No B422457 depicts a typical example. The two handles between the wheels operate the underfloor hopper doors and at the far right the handbrake lever can be seen (small BR type). BR built both non-fitted and vacuum brake fitted examples, however today the majority of those in traffic are vacuum fitted and the handful of non-fitted examples are scheduled for early withdrawal. Livery of fitted 21 tonne coal hoppers should be bauxite, whereas unfitted examples should be light grey, although this can no longer be relied upon for total accuracy. *Colin J. Marsden*

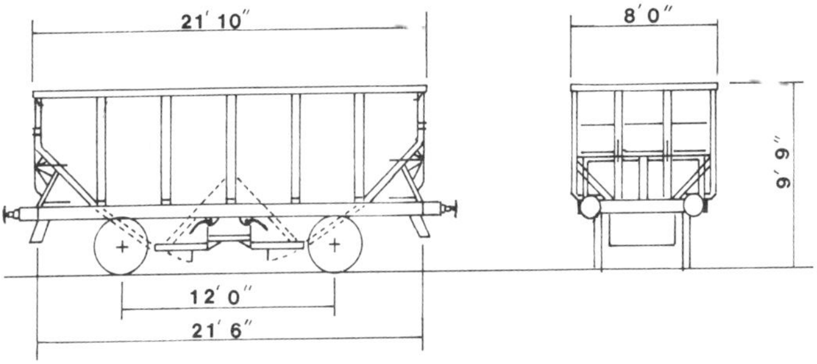

HTO, HTV
Tare: 9.5 tonne
Carrying capacity: 21.5 tonne
Cubic capacity: 23cu m
Minimum curve: 26m
Body: Steel

Group I: International Ferry Wagons (Other Railways Administrations)

Code	*Description*	*GLW Range*	*Brake Type*
ICA	Tank (2 axle or bogie)	20-50t+	Air
ICB	Tank (2 axle or bogie)	20-50t+	Air brake, vacuum piped
ICP	Tank 2 axle	30-39t	Vacuum piped
ICQ	Tank 2 axle	20-29t	Air piped
ICR	Tank 2 axle	20-39t	Dual piped
ICW	Tank 2 axle	30-39t	Vacuum brake, air piped
ICX	Tank (2 axle or bogie)	20-50t+	Dual
IFB	Flat	15-28.5t	Air brake, vacuum piped
IFR	Flat	21-45.5t	Dual piped
IFX	Flat	27-27.5t	Dual braked
IIB	Refrigerated van	13-25.6t	Air braked, vacuum piped
IIX	Refrigerated van	13-25.6t	Dual braked
ILB	Large van	21.5-26t	Air braked, vacuum piped
ILX	Large van	24.5-26t	Dual braked
IMB	Medium van	2-27t	Air braked, vacuum piped
IMX	Medium van	19-26t	Dual braked
INX	Small luggage van	7t	Dual braked
IOB	Open high	27.5-29.5t	Air braked, vacuum piped
IPB	Privately owned ferry vehicle	10-110t	Air braked, vacuum piped
IPX	Privately owned ferry vehicle	10-30t	Dual braked
ISB	Small van	23t	Air braked, vacuum piped
ISX	Small van	20-23t	Dual braked
ITX	Transfesa (flat or van)	25-29t	Dual
IXB	Special bogie (plate or well)	33-60t	Air braked, vacuum piped

Vehicles classified under Group I in the TOPS system are Continental vehicles owned by foreign railways or private companies that conform to the UIC standard and able to operate on British tracks within BR train formations. All are fitted with anchor points for stowage on train ferries.

Right: Painted wagon label of Group I vehicle. This ferry van is owned by SNCF — the French National Railways, and its number is 238 0 314-2. The code letters Hbfs are internal coding for the SNCF and have no relevance to British operation. The owning railway is always given under the RIV sign by the relevant code letters. See table later in section. *Colin J. Marsden*

Above: ICB. BR code ICB is described in the BR TOPS manual as a tank wagon conforming to ferry wagon requirements, owned by a foreign administration, fitted with air brake equipment (in the case of the wagon illustrated the single pipe system) and a through vacuum pipe (not vacuum brake fitted). Unlike British vehicles that have a separate TOPS code for different weights and wheel arrangements, all foreign tank wagons operable on BR are classified as IC with only the brake fitting differing. The bogie vehicle shown with a loading capability of 57.0 tonne, is operated by Rhone Poulenic from a base on the SNCF. *Colin J. Marsden*

Below: ICB. This interesting tank wagon with a GLW of 29.9 tonne is operated from the Swiss Railway SBB, and is owned by Ciba-Geigy of Basle, Switzerland, operating regularly between Basle, Zurich and England via the Harwich-Zeebrugge train ferry. This vehicle is again air braked with a through vacuum pipe. Note the interesting verandah at the near end where the handbrake wheel is positioned. No 739 0 539-0 was photographed at Wittlesford. *Colin J. Marsden*

Above: IFB. Continental railway owned flat wagons come under the TOPS classification of IF; various differing types come under the same classification with carrying capacities varying between 15 and 25 tonne. The same classification applies to both two axle and bogie vehicles. As with the previous International types, various brake fittings prevail. IFB — air braked with vacuum through pipe, IFR — non brake fitted with both air and vacuum through pipes, and IFX — dual brake fitted. IFB No 414 5 216-1 owned by the SNCF is shown here acting as a barrier between two nuclear flasks. *Colin J. Marsden*

Below: IIB. A number of refrigerated vans operate between the Continent and England, transporting perishable foodstuffs such as fruit or meat. The TOPS code II covers these vehicles of which there are several types — small, medium, large, fitted with meat hooks, and Inter-Frigo vehicles, varying between 13 and 25 tonne. Within the group there are two brake types: IIB — air braked vacuum through pipes, and IIX — dual braked. The vehicle illustrated conforms to TOPS code IIB and is No 808 9 849-4 owned by the Italian Railway. *Michael J. Collins*

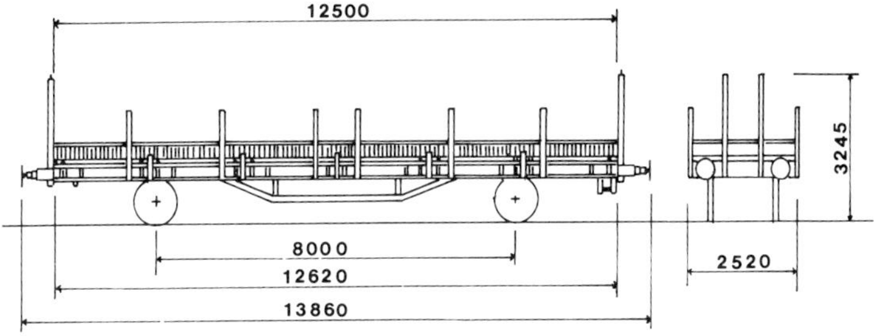

IFB, IFX
Tare: 12.5 tonne
Carrying capacity: 27.5 tonne
Floor area: 27.7sq m
Minimum curve: 60m
Body: Steel
Floor: Wood

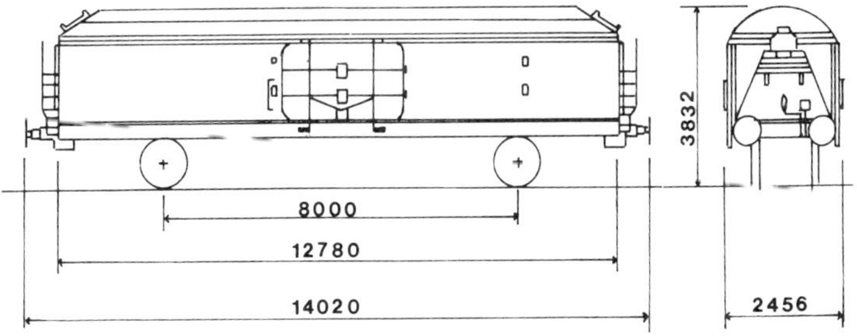

IIB
Tare: 14.4 tonne
Carrying capacity: 25.6-25.9 tonne
Cubic capacity: 43.9cu m
Minimum curve: 60m
Body: Wood, thermal block insulation
Floor: Wood

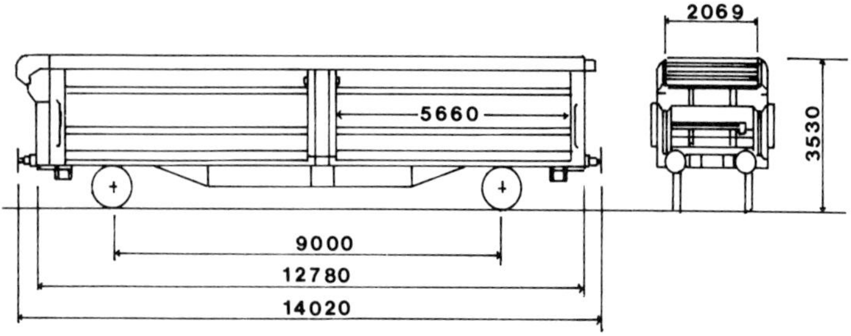

ILB
Tare: 14.5 tonne
Carrying capacity: 25.5 tonne
Cubic capacity: 61.2cu m
Minimum curve: 70m
Body: Aluminium
Floor: Wood

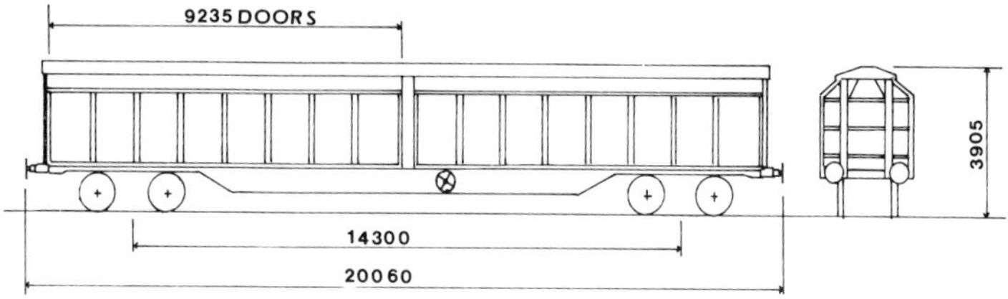

ILB
Tare: 26 tonne
Carrying capacity: 54 tonne
Minimum curve: 60m
Body: Metal (steel/aluminium)
Floor: Wood

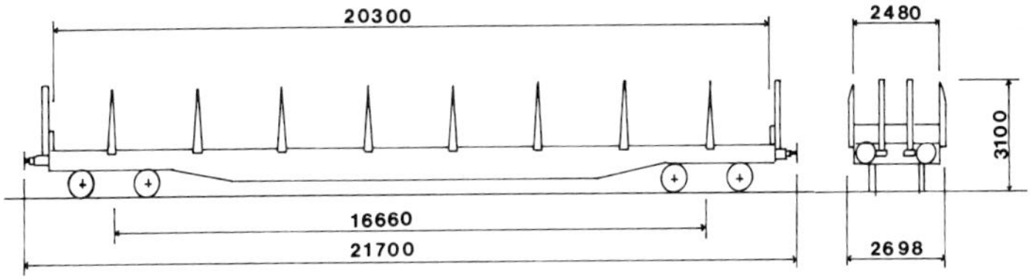

IPB
Tare: 23 tonne
Carrying capacity: 57 tonne
Floor area: 50.3sq m
Minimum curve: 35m
Body: Metal
Floor: Wood

International Wagon Identification Codes

Code	*Country*	*Initials*
20	Soviet Union	SZD
50	East Germany	DR
51	Poland	PKP
52	Bulgaria	BDZ
53	Rumania	CFR
54	Czechoslovakia	CSD
55	Hungary	MAV
63	BLS-Switzerland	BLS
71	Spain	RENFE
72	Jugoslavia	JZ
73	Greece	CH
74	Sweden	SJ
75	Turkey	TCDD
76	Norway	NSB
80	West Germany	DB
81	Austria	OBB
82	Luxembourg	CFL
83	Italy	FS
84	Netherlands	NS
85	SBB-Switzerland	SBB
86	Denmark	DSB
87	France	SNCF
88	Belgium	SNCB

Above: IMB, IMX. The majority of International van stock that visits the BR network conforms to the TOPS code IMB or IMX (International medium van, either fitted with air brakes and a through vacuum pipe, or dual brake equipment). This type of vehicle can weigh anything between 19 and 26 tonne, and normally has one large sliding door mid-way along the body, with four opening top lights (sliding type). SNCF owned 26 tonne capacity No 230 0 314-2 is shown here. *Colin J. Marsden*

Left: IMB, IMX. End detail of continental owned ferry van. Ventilation grilles at the top, standard buffing gear at the bottom, with screw coupling, the through vacuum pipe is to the left of the coupling and the air brake pipe to the right. In common with most European stock this vehicle operates on the single pipe air system — no main reservior pipe. *Colin J. Marsden*

Above: IOB. Open or open high vehicles owned by foreign railway operators are classified as IO. Vehicles conforming to brake type B only are currently operating on to BR tracks. They can have a carrying capacity of between 27.5 and 29.5 tonne and normally have at least two large opening doors on each side, as shown here on French operated (SNCF) IOB No 609 4 040-1 seen at Toton, painted in bauxite livery with white lettering.
Colin J. Marsden

Below: Group P within the International (I) TOPS code is allocated to privately owned vehicles authorised to operate on both foreign and BR tracks. This code covers all types of vehicle from the smallest flat wagon to the largest bogie hopper, as illustrated here. The largest bogie hopper wagon (in size), to operate on BR tracks is this type of powder hopper capable of taking up to 58 tonne of cargo. All are air brake fitted with through vacuum pipes and either owned by *Polybulk* or *Cita*, and normally operate in block trains of up to 10 vehicles. *Cita* vehicle No 099 8 072-3 operating from the Belgian SNCB network is shown here. The interior of these large vehicles is divided into three sections, marked on the outside by the letters A, B and C. The discharge of the appropriate compartment is achieved by turning the relevant underframe mounted wheel. *Michael J. Collins*

Above: IPB. A large number of German operated *Cargowaggon* type vans operate on the BR network. These bogie vans are capable of carrying up to 54 tonne of cargo and have three large openable side positions to assist with rapid loading and unloading, giving access to one-third of the vehicle at a time. This van design has a length of 19.86m over headstocks, and a total height of 3.91m with a loading height of 2.26m. *Cargowaggon* No 279 7 502-2 with its home base on the DB in West Germany is illustrated here. *Colin J. Marsden*

Below: IPB. In addition to the *Cargowaggon* vans that operate on the international circuit, many flat bogie vehicles are used between West Germany and England. The vehicle illustrated here carrying raw cut hardwood is a 20.3m long vehicle mounted on two four-wheeled bogies, and also has removable side support stanchions. *John C. Baker*

Right: IPB. End details of *Cita* or *Polybulk* type large hopper wagon of IPB type. On the buffer beam to the left of the coupling a through vacuum pipe can be seen, and to the right a main reservoir and air brake pipe. No lamp iron is provided on this vehicle so the tail lamp is hung on the coupling hook. *Michael J. Collins*

Above: ITX. A sizeable fleet of *Transfesa* vehicles are operated by the Spanish Railways RENFE. Three distinct types come to England — 30-40ft flat wagons, 25-26 tonne medium vans and 26-27 tonne medium vans fitted with end doors. These vehicles are used almost exclusively for transporting fruit and vegetables, all are dual braked to the single pipe air system, and painted blue. The illustration shows a 26.5 tonne van with end doors, No 028 5 872-4, at Spalding. *Michael J. Collins*

Below: IXB. The letter X in the TOPS wagon type code system normally refers to exceptional or very special vehicles, and in the case of international wagons this is no exception. This large bogie wagon with no less than 16 wheels is a nuclear flask and is capable of carrying up to 97.5 tonne. The vehicle is owned by a French operator and its normal home base is on the SNCF system somewhere in the Paris area. No 998 5 001-5 was photographed departing from the Windscale Nuclear establishment at Sellafield, bound for a dockyard and return to France. The whereabouts and further information on these vehicles is of a classified nature. *Colin J. Marsden*

Group J: Steel Carrying — Bogie Coil

Code	*Description*	*Capacity*	*Brake Type*	*Remarks*
JAV	Bogie strip coil	42.5t	Vacuum	
JEV	Bogie strip coil 'E' (with nylon hood)	44.5t	Vacuum	Ex warflat
JGV	Bogie strip coil 'G' (with nylon hood)	44.5t	Vacuum	Ex warflat
JIX	Bogie strip coil — RIV (with nylon hood)	53.5t	Dual	
JKX	Bogie strip coil 'K' (with nylon hood)	61t	Dual	
JMV	Bogie strip coil 'M'	42.5t	Vacuum	Ex bogie bolster D
JPV	Bogie strip coil 'P'	32.5t	Vacuum	Ex bogie bolster E
JRV	Bogie strip coil 'R'	32.5t	Vacuum	Ex bogie bolster E
JTV	Bogie strip coil 'T' (with nylon hood)	61t	Vacuum	Ex slab
JUV	Bogie hot rolled coil 'U'	32.5t	Vacuum	Ex bogie bolster E
JVV	Bogie strip coil 'V' (with nylon hood)	61t	Vacuum	
JYV	Bogie strip hot coil	42.5t	Vacuum	Ex bogie bolster D
JZV	Bogie slab coil	46t	Vacuum	

Below: JGV. Vehicles classified under Group J of the TOPS system are bogie steel coil carriers. When manufactured the steel is often rolled into coils for ease in transportation and for this usage specially adapted wagons have been introduced. These wagons operate on specific duties in block formations. JGV No W160134 is a bogie strip *coil G* fitted with two nylon hoods to protect its load, this vehicle was converted from a Warflat built prior to nationalisation and has a maximum load of 44.5 tonne. *Paul Shannon*

Above: JPV. Seventy-five members of the Bogie Bolster E (BEV) type were converted during the 1970s to *Bogie Strip Coil Ps* by forming a longitudinal cradle to carry steel coil down the centre of the wagon, by either utilising the former bolsters or other recovered material from scrapped wagons; otherwise the vehicles remained as bogie bolster Es and still carry numbers from that fleet. JPV No B924813 is shown at Whitemoor.
Michael Rhodes

Below: JTV. This view of a JTV bogie strip *coil T*, fitted with a nylon hood, clearly shows the arrangement under the hood where the steel coils rest in wells formed into the body of the wagon and supported by additional side members. This type of wagon has a carrying capacity of 61 tonne and is fitted with vacuum brake equipment; handbrakes are fitted to both ends of the wagon. No B949522 is shown at Gartosh BSC works in Lanarkshire.
British Railways

Right: JUV. A further conversion of the Bogie Bolster E (BEV) to steel carrying use, came during the early 1970s when a small fleet of 20 were rebuilt as *Bogie Coil U*. These vehicles were equipped to accept their new cargo of hot rolled coils, having new steel floors formed of perforated steel channels and mesh for heat dissipation, side curb rails were retained and five central coil anchorage points were provided. JUV No B923865. *Michael Rhodes*

Centre right: JVV. A fleet of *Bogie strip coil V* vehicles were introduced during 1957 with a carrying capacity of 61 tonne. Each vehicle has two separate nylon covered cradles to take the strip coil, all vehicles are fitted with vacuum push brakes. They are normally confined to operation in South Wales and No B949079 is illustrated here at Radyr. *Michael Rhodes*

Below right: JZV. In 1962 the BR workshops at Swindon produced under lot number 3424 a fleet of 58 Slab/Coil wagons now classified as JZV. Vehicles No B94551-949608 were intended for use on South Wales steel duties and bore painted instructions for their return to WR. All are fitted with vacuum brakes and roller bearing axles, and have a maximum load of 46 tonne. *British Railways*

Group K: Steel Carrying — 2 Axle Coil

Code	*Description*	*Capacity*	*Brake Type*	*Remarks*
KAV	Strip Coil A (with nylon hood)	21.5t	Vacuum	Ex mineral
KBV	Strip Coil B (with nylon hood)	24.5t	Vacuum	
KCO	Strip Coil C	24.5t	Unfitted	Ex pig iron
KDV	Strip Coil D	24.5t	Vacuum	Ex plate
KEV	Strip Coil E	21t	Vacuum	Ex plate
KGO	Pig Coil	30.5t	Unfitted	Ex pig iron
KJO	Strip Coil J	24.5t	Unfitted	Ex tippler
KLV	Strip Coil L	22.5t	Vacuum	Ex shochood B
KNO	Strip Coil	13t	Unfitted	Ex single bolster
KRV	Strip Coil R	24.5t	Vacuum	Ex plate
KSV	Strip Coil S	13t	Vacuum	Ex high
KTA	Strip Coil T	31.5t	Air	
KYV	Hybarcoil	13t	Vacuum	Ex hybar

Below: KDV. Steel carrying wagons mounted on two axle frames are classified under TOPS as Group K. Again these are usually found operating in block formations in major steelworking areas. This view of a KDV *Coil D*, converted from a 22 tonne plate wagon, has been included to show the storage arrangement for the strip steel coil; the wagon carries three demountable cradles, each able to carry 7 tonne of steel, and supplied with its own hood. Vehicles forming this group were first converted from 1964 when 71 wagons were introduced for use on the Parkgate, Rawmarsh (Rotherham) and Great Bridge route.
British Railways

Above: KEV. Another type of coil steel carrying vehicle adapted from plate wagons are these *Coil E* or KEV examples, which are similar to the *Coil R* (KRV) wagons described later. This fleet has only been converted in recent years and retain most of the plate wagon features, with main alterations being the extension of the corner stanchions upwards by 18in, and two additional central end stanchions. There are nearly 200 of these vehicles in existence and all operate from the Cardiff area. *Colin J. Marsden*

Below: KJO. A number of short sided tippler wagons introduced during 1958, not fitted with brake equipment, were converted into *Strip Coil J* vehicles during the 1970s and usually can be seen operating in South Wales. These vehicles are fitted with wooden internal supports to accept the steel coils and can carry up to 24.5 tonne. KJO No B386852, painted in unfitted grey livery, is shown here with Hybox type axle boxes. *Michael Rhodes*

Above: KRV. TOPS code KRV is given to former plate wagons converted to strip *Coil R* vehicles. This plate wagon rebuild carries coiled steel rod from steel works in South Wales and Scunthorpe areas. Apart from the underframe, none of the former plate wagon remains, the ends being replaced by four stanchions with a plank base, and the sides now have five support stanchions holding an angle iron, together with four floor battens to support the steel coil. All vehicles are fitted with vacuum brakes, KRV No B934891 is seen here in the Cardiff area. *Colin J. Marsden*

Below: KYV. A fleet of former *Hybar* wagons were converted to *Hybarcoil* during the early 1970s for the conveyance of steel coil. From the outside these vehicles look as though they are a standard OWV five-plank wagon with a tarpaulin bar, and therefore recognition of this type is best achieved by reference to the TOPS code. Supports are provided inside to take the coils. Vehicles are fitted with push type vacuum brakes and can carry a 13 tonne load. No B491890 is illustrated here at Ebbw Junction, Newport. *Michael Rhodes*

Group M: Mineral Wagons (Not Hopper Fitted)

Code	*Description*	*Capacity*	*Brake Type*	*Remarks*
MCO	Mineral (side and end doors)	16.5t	Unfitted	Some fitted with bottom doors
MCV	Mineral (side and end doors)	16.5t	Vacuum	Some have clasp style brakes
MDO	Mineral (side doors)	21.5-25t	Unfitted	Some vehicles have end doors
MDV	Mineral (side doors)	21.5t-25t	Vacuum	Some vehicles have end doors
MFA	Mineral, ferrous scrap	32.5t	Air	
MSO	Tippler	26.5-27.5t	Unfitted	
MSV	Tippler	26.5t	Vacuum	
MTV	Tippler	24t	Vacuum	
MXV	Mineral (side and end doors)	16.5t	Vacuum	Fitted with push brakes

Below: MCO. The most numerous wagon ever built for BR freight operation is the 16 ton (now 16.5 tonne) mineral wagon. Since nationalisation alone nearly 250,000 have been constructed, however today only a handful survive. Both unfitted and fitted examples were built and now classified as MCO (unfitted) and MCV (Vacuum fitted clasp brakes). MCO No B564140 built in 1955 is shown. *Colin J. Marsden*

Left: A long loose coupled train formed of a mixture of Hopper (Group H) and Mineral (Group M) vehicles. Behind the Class 47 is an HTV followed by five MDVs and a mixture of HTV, MCV, MCO vehicles. *Colin J. Marsden*

Above: MCV. The design of the 16 tonne mineral wagon dates back to pre-nationalisation LMS days, however post-nationalisation built vehicles have a simplified brake layout. All vehicles have an opening central panel on each side, and the majority have an openable end, indicated by a white diagonal painted on the side (on left in illustration). The majority of remaining vehicles have welded bodies but a few riveted examples still survive. Some earlier built members have a small openable section above the side doors. Initially these wagons were intended for coal conveyance, but over the years the vehicles have been put on a variety of duties including scrap steel, sand, ballast etc. B552572 is shown here. The MCV TOPS code is applied to 16.5 tonne mineral wagons fitted with clasp type brakes (two brake blocks, one each side of each wheel); vehicles fitted with push brakes (one block per wheel) are classified as MXV. *Colin J. Marsden*

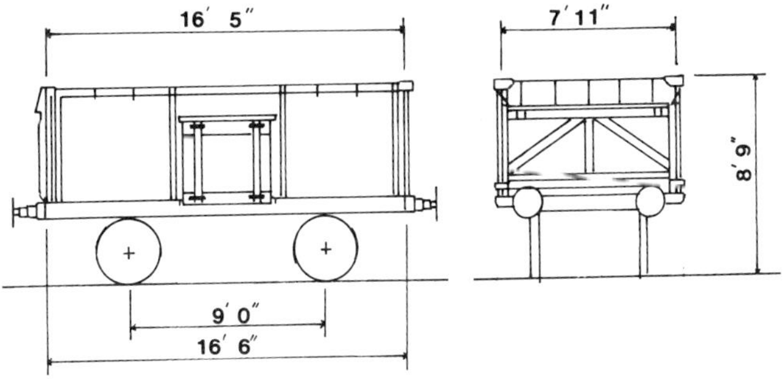

MCO, MCV
Tare: 7.7 tonne
Carrying capacity: 16.5 tonne
Cubic capacity: 18.3cu m
Minimum curve: 21m
Body: Steel
Floor: Steel

MDO, MDV, MXV

Above: MDO, MDV. The GWR pioneered the use of high capacity coal or mineral wagons back in the 1930s, building a considerable number and placing them on loan to a number of South Wales operators. The design continued into BR days and a number of 21 tonne vehicles of both fitted and unfitted type, were constructed, also for work in the South Wales area. Several detail differences exist within the fleet such as welded or riveted bodies, side door design, whether fitted with oil or roller bearing axles. MDO No B316573 (fitted with roller bearing axles) stands at Radyr with the end door nearest the camera. This type of wagon now operates throughout the country. MDV type is identical but not illustrated. *Colin J. Marsden*

Below: MXV. The TOPS code MXV is a recent addition for standard 16.5 tonne mineral wagons of the MC type, fitted with push type vacuum brakes (brake block applied to one side of the wheel only) rather than the clasp type. As this TOPS code has been introduced since the commencement of the standard wagon identification system there are some MXVs to be seen still carrying the code MCV. MXV No B574829 is seen here with the adjoining vehicle still bearing the MCV code although it has push brakes. *John Tuffs*

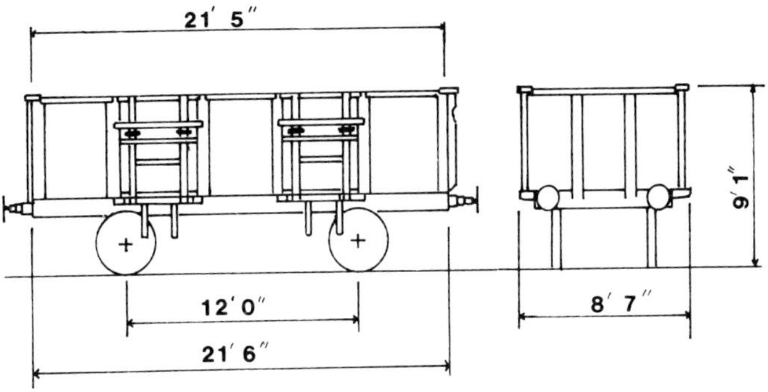

MDO, MDV
Tare: 10.5 tonne
Carrying capacity: 21.5-25.5 tonne
Cubic capacity: 27.2cu m
Minimum curve: 26m
Body: Steel
Floor: Steel

Below: MSO, MSV. Between 1951 and 1962 a fleet of 26 ton (now 26.5 tonne) Iron Ore tippler wagons were constructed resembling the 16 ton mineral wagon in almost every respect, except that no doors were provided and the maximum load was raised by some 10 tons. Today these vehicles are used for stone conveyance usually in block train formations. Non-fitted and vacuum braked vehicles were constructed, however those now remaining are mainly vacuum brake fitted. No B386019 built at Derby works is shown here carrying block set No 7603 above the wagon information panel. A special feature of these wagons is the facility for being unloaded by mechanical grab or by tipping.
Colin J. Marsden

Group O: Open Wagons

Code	*Description*	*Capacity*	*Brake Type*	*Remarks*
OAA	Open	31.5t	Air	
OBA	Open (with Turnover bolster)	30.5-31.5t	Air	
OCA	Open (with bolster)	31.5t	Air	
OHB	High — steel sided	13t	Air, vacuum piped	
OHV	High — steel sided	13t	Vacuum	Some vehicles of the hybar type
OIX	Tube (Continental Std)	26.5t	Dual	
OJX	High	21.5t	Dual	Ex ferry
OLV	Low	13t	Vacuum	
OUV	Shochood B	20.5t	Vacuum	
OWV	High/Hybar/Hyhood	13t	Vacuum	Some vehicles have wooden sides, while others have drop sides

Below: OAA. The OAA series of wagons were the first of the modern purpose built air braked vehicles for the *Railfreight* system of today. They were introduced from 1971 and numbered in the 100000-100099 series. When built these vehicles were called *OPEN AB*, as the TOPS wagon system had then not been introduced. These vehicles are designed to convey general merchandise or palletised traffic and have three drop side doors on each side, and if the door spacer stanchions are removed, full length access can be gained. No 100027 is seen here when new. *British Railways*

Above: OBA. This fleet of 31 tonne open wagons carrying numbers 110001-110800 were introduced between 1977-1979 for the conveyance of steel, general merchandise and palletised loads, they are fitted with full length drop doors in four sections separated by removable telescopic stanchions; there are also six turnover bolsters on the floor. When in tare condition the OBA weighs 14.5 tonne, and when loaded 32 tonne for 60mph operation and 27 tonne for 75mph running. Bauxite brown No 110668 with *Railfreight* insignia is photographed here. These wagons were the development of the OAA type. *Colin J. Marsden*

Below: OBA, ZDA. During 1982 480 of the OBA type were transferred to the departmental fleet for use by the Civil Engineering Departments. Upon transfer the vehicles were recoded as ZDA and named *Bass*. The *Railfreight* name on the sides was painted out and the number prefixed with the letters DC, and some vehicles received a pool number. *Bass* No 110151 is seen here at Woking. *Colin J. Marsden*

OBA, OCA

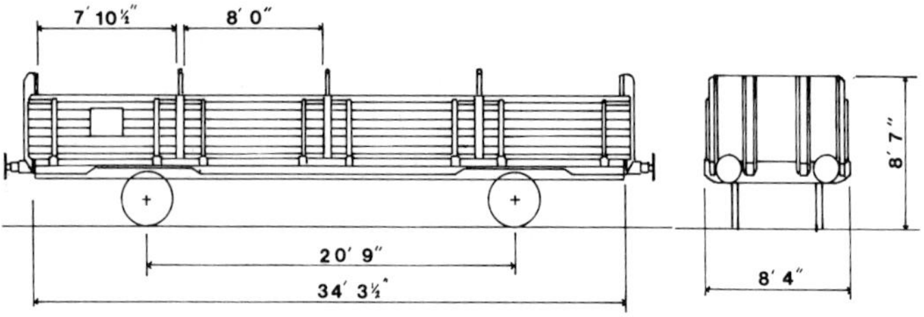

OBA
Tare: 14.5 tonne
Carrying capacity: 32 tonne at 60mph
27 tonne at 75mph
Cubic capacity: 22.6cu m
Minimum curve: 41m
Body: Fixed steel ends, timber lined full length wooden drop doors
Floor: Wood with six turnover bolsters

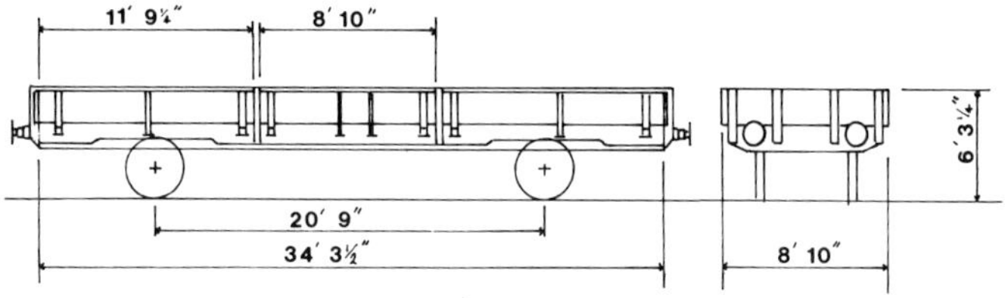

OCA
Tare: 14 tonne
Carrying capacity: 31 tonne at 60mph
26 tonne at 75mph
Cubic capacity: 23.2cu m
Minimum curve: 41m
Body: Fixed steel ends, timber lined full length steel drop doors
Floor: Wood with six turnover bolsters

Above: OCA. The latest type of *Railfreight* open wagon emerged from 1979 classified as OCA. This fleet is of all steel construction, unlike the two previous types, and is again designed to convey steel, general merchandise and palletised traffic. Three dropside doors with removable support stanchions are provided giving access to the complete wagon length if required; there are six turnover bolsters on a wooden floor. The wagon can be loaded up to 31 tonne for 60mph operation and 26 tonne for 75mph running. No 112317 is illustrated painted in *Railfreight* red livery. *Colin J. Marsden*

Below: OCA. This view of an OCA wagon clearly shows the end detail with standard screw shackle coupling and dual air pipes on the buffer beam. There is a lamp iron provided on the left of the two end supports with tying points on the end bodywork. The centre side door with an opening of 8ft 11in is in the open position. No 112067 is shown here at Hartlepool during a wagon exhibition. *Colin J. Marsden*

Above: OHV. The standard 13 tonne high sided open wagon with a steel body, is classified under TOPS as OHV. The type owes its design to the LNER who built a number during the 1940s for its *Green Arrow* fast freight service. After nationalisation BR continued the build with similar wagons, some of the later examples having a more shallow body. All are fitted with vacuum brakes. LNER built No E281096 with clasp brakes, is illustrated here at Wimbledon depot where it arrived loaded with brake blocks. *Colin J. Marsden*

Below: OHV. BR built OHV No B496377 illustrated here, was introduced during 1955, one of 650 fitted with push type brakes and a slightly more shallow body. The vehicle was photographed at BREL Doncaster works containing a number of items for repair from a diesel depot. When this photograph was taken in 1982 it carried no TOPS code.
Colin J. Marsden

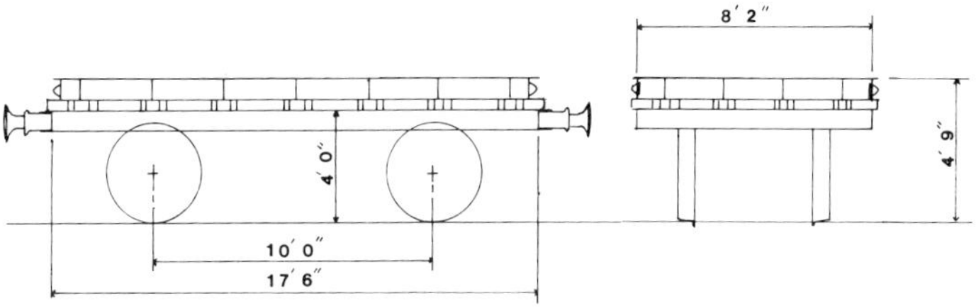

OLV
Tare: 6.5 tonne
Carrying capacity: 13 tonne
Minimum curve: 21m
Body (sides): Steel
Floor: Wood

Below: OWV. The wooden bodied variant of the standard 13 tonne open wagon is classified as OWV. It is often called a five plank and was a development of an LMS design. During the 1940s many LMS five plank wagons received damaged ends caused by shifting loads and to overcome this problem corrugated steel ends were adopted as standard on subsequent builds. No B492204 is shown here with push type brakes. Some of these vehicles have tarpaulin bars. *Colin J. Marsden*

Group R: Railway Operating Vehicles

Code	*Description*	*Brake Type*	*Remarks*
RBA	Barrier wagon	Air	Ex steel AB
RBB	Barrier wagon	Air, vacuum piped	Ex steel AB or tube
RBV	Barrier wagon	Vacuum	Ex ale pallet, banana or fish van
RBX	Barrier wagon	Dual	Ex carfit, high van, or motorcar van
RFO	Adaptor wagon for LTE stock	Unfitted	
RFQ	Adaptor wagon for FLT stock	Air piped	
RFQ	Adaptor wagon	Air piped	Ex tippler
RFV	Adaptor wagon for LTE stock	Vacuum	
RRA	Runner wagon	Air	
RRB	Runner wagon	Air, vacuum piped	
RRV	Runner wagon	Vacuum	
RRW	Runner wagon	Vacuum, air piped	Ex lowmac
RTV	Brake tender	Vacuum	

The TOPS Group R covers non-revenue earning vehicles operated for the railway's own benefit and classified as Railway operating vehicles. There are a number of recent additions to this group, and one off examples, and it is not possible to include all of them. Vehicles of this group are usually converted from standard freight stock.

Below: The RB group is for barrier vehicles, this term is applied to any vehicle used to protect another, or coupled between a locomotive and load. This could be for reasons of spacing, protection from a dangerous load, or distributing weight. The barrier shown here was formerly a 22 tonne *Palvan* introduced in 1963. The livery of vehicle No B787094 was dark blue when photographed at Plymouth Friary in 1982. An interesting point regarding this vehicle is that although dual brake equipment is fitted on the buffer beams, it is coded RBV and labelled 'vac brake only'. Either the air pipes should be removed or the vehicle recoded to ensure that it is not mis-used. *Colin J. Marsden*

Above: RBV. This barrier vehicle, built in 1960 by BR at Darlington Works was formerly a 12 tonne special parcels van (SPV) and before that an insulated fish van to BR passenger train standards. The vehicle has a long wheelbase and is fitted with clasp type vacuum brakes, and GWR style roof and end detail. No W87964 was photographed at Severn Beach in 1982. *Colin J. Marsden*

Below: RBX. A number of various designs of flat wagon have been drafted into railway operating department stock and are often used as reach or spacer vehicles for exceptional loads or barriers between locomotives and oil or other dangerous goods. No B748136 a former Carfit C is illustrated here. This vehicle was constructed to RIV standards, and a similar vehicle, in its correct application, can be seen in the Group F section. *John Tuffs*

Above: RBX. A considerable number of British owned continental ferry vans fitted with dual brake equipment have been transferred to railway operating department use as barrier vehicles; these are long wheelbase vans and especially suited to their new role. No B786931 illustrated was built in 1962 as a ferry vanfit conforming to RIV standards and operating with the single pipe air system. In later years of revenue earning service the vehicle was part of the *Railfreight* fleet. Photographed at Severn Beach. *Colin J. Marsden*

Centre left: RBX. This dual braked barrier conforming to international RIV standards was formerly a 14 ton SR ferry van built in 1958 for the conveyance of vehicles between England and the Continent; it was based on the SR designed CCT. This vehicle was originally numbered in the International numbering system but now only carries BR number and is painted in bauxite livery with white handrails and lettering. B889009 was photographed at Derby during 1982. *John Tuffs*

Bottom left: RFQ. When the inner vehicles of freightliner sets require individual movement, usually at a freightliner maintenance depot, they cannot be coupled to normal stock because of their bar couplings. Therefore a fleet of adaptor wagons was built with bar couplings at one end and dual shackle and bar drawgear at the other; vehicles are unbraked but have a through air pipe. These vehicles were introduced during the 1960s and were converted from *Palbrick B* wagons. No B462783 is seen here at Old Oak Common depot. *Colin J. Marsden*

Above: RFO. The couplings on London Transport rolling stock are in no way compatible to BR vehicles or locomotives and therefore a small number of adaptor wagons are in service. The need to haul LT stock by a BR locomotive is infrequent and usually only occurs when vehicles are en route to scrap dealers. In this picture we see a Class 25 No 25.143 hauling a train of R stock along the Midland main line bound for Booths scrap dealers at Rotherham. The RFO BR/LTE adaptor wagon, converted from a five plank wagon of OW group, is coupled between the locomotive and LTE stock. *Ian Bald*

Right: RRA. The railway internal code RR is allocated to runner vehicles, these are wagons used as spacers within a train for reason of load, weight dissipation or size. No 400254 illustrated is a disc air brake fitted example converted from a Steel AB (SAA) which was introduced in 1971. It is normal practice for railway operating department vehicles to be branded 'Not for Loading', however the legend cannot be seen on the vehicle photographed at Woking. *Colin J. Marsden*

Group S: Steel Carrying — 2 axle (Excluding Coil)

Code	*Description*	*Capacity*	*Brake Type*	*Remarks*
SAA	Steel AB	31.5t	Air	
SAB	Steel AB	31.5t	Air, vacuum piped	
SBA	Steel ABB	31.5t	Air	
SEA	Steel AB	30t	Air	
SMO	Ingot mould	27.5t	Unfitted	Ex tippler
SOV	Pipe	12t	Vacuum	
SPA	Plate/Ingot	31-31.5t	Air	
SPV	Plate	22.5t	Vacuum	
STV	Tube	20.5-22.5t	Vacuum	Some vehicles have side doors while others have drop or fixed sides
SWV	Twin bolster	13t	Vacuum	

Below: SAA, SAB. Group S is the third group within the TOPS system used for steel carrying wagons, two axle vehicles intended for conveying steel in other than coils. However some vehicles lend themselves to coil transportation and other applications. The SAA, SAB fleet was introduced from 1971 when 300 were constructed in the number range 400000-400299. These wagons are normally found operating in block formations especially from Scotland and South Wales. Some of these vehicles have been taken into railway operating department stock as runners. SAA No 400048 is illustrated here. *British Railways*

Above: SOV. Following on from LNER design, BR produced over 1,900 12ft WB 12 tonne *Pipe* wagons from 1969, numbered in the ranges B484150-484199 and B740000-741949. All were vacuum brake fitted, some having push, and others clasp style brakes. Vehicles were built to transport steel piping mainly from Stanton & Staveley Ironworks, but over the years they have seen far wider use as general merchandise carriers. In this view No B741103 stands outside Doncaster Works. *Colin J. Marsden*

Below: SPA. This type of modern air braked steel wagon began appearing from BREL works during the late 1970s and to date some 1,600 have been introduced in the number range 460000-461601. Vehicles have a wheelbase of 20ft 9in with three side opening doors, giving uninterrupted access to the 34ft 3in inside length with removal of the centre stanchion. These versatile wagons were designed for the movement of steel plate, billets, bars and other steel traffic. Floors are of steel with square wooden cross battens at regular intervals to support the load which is secured by hand operated ratchet tensioners. SPA No 460710 in *Railfreight* red livery is shown here. *Colin J. Marsden*

Above: Wagon label of modern air braked vehicle of SPA type, with TOPS code at the top, maximum load at middle left, tare weight at middle right, and wagon number below. On modern air braked stock plates are painted white. *Colin J. Marsden*

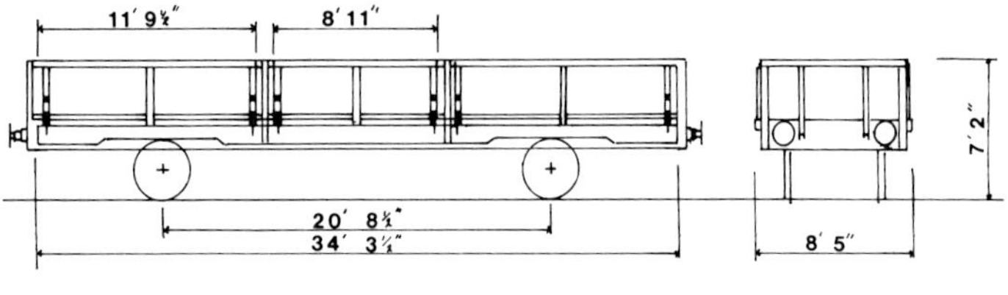

SPA
Tare: 14.5 tonne
Carrying capacity: 31.5 tonne
Minimum curve: 41m
Body: Fixed steel ends, steel drop sides
Floor: Steel, wooden battens

bove: SPV. Shortly after nationalisation a large number of 22 tonne unfitted and vacuum
raked flat *Plate* wagons were constructed, mainly for the transportation of plate steel, but
ould also accommodate other loads such as motor cars and industrial machinery. Today
o unfitted vehicles remain in service and the vacuum fleet is fast diminishing. A number of
ninor variants exist within the ranks of remaining vehicles such as axle box design, and
tyle of buffers. No B934816 shown here was introduced during 1960 and has *Hybox* axle
oxes and Oleo pneumatic buffers. *Colin J. Marsden*

elow: STV. For the transportation of steel tubes from steelworks to industrial sites a
izeable fleet of *Tube* wagons based on LMS/LNER designs was built by BR soon after
ationalisation. Under the TOPS system these vehicles became STV. Four types of vehicle
xist with ie: 1. drop sides in two sections with extended height ends, 2. drop sides in two
ections with standard height ends, 3. metal ended vehicles with a small central door, 4.
olid sided vehicle with metal ends. No B730053 illustrated here has two half length drop
ides and extended ends. *Colin J. Marsden*

Above: STV. The most common *Tube* wagon seen today is the BR standard vehicle based on an LMS design with a small central drop door and fitted steel ends to avoid damage by shifting loads. These vehicles can accommodate 22.5 tonne of merchandise. No B730719 illustrated was built at Derby in 1960 and has *Hybox* style axle boxes. *John Tuffs*

Below: STV. Some of the *Tube* fleet bear the name *Batten* or *Tube/Batten*, this applies to vehicles that have wooden slats transversely fitted to the floor to raise the load. Labelled as a *Batten* and fitted with roller bearing axles No B730747 stands at Woking loaded with concrete cable troughing. These vehicles are mainly used for the transporting of general merchandise and seldom seen loaded with tube. *Colin J. Marsden*

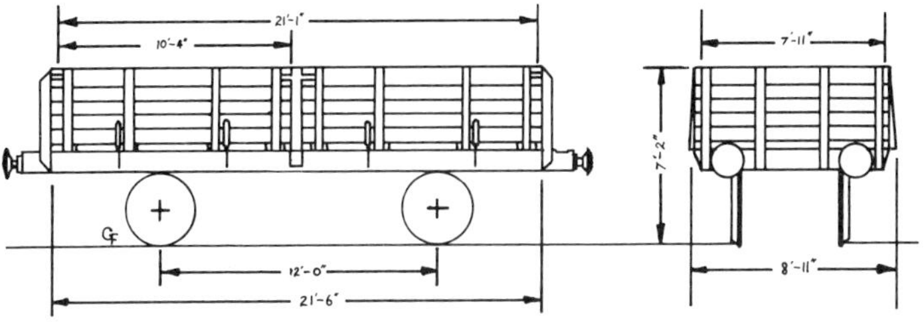

SOV
Tare: 8.5 tonne
Carrying capacity: 12 tonne
Cubic capacity: 15.34cu m
Minimum curve: 26m
Body: Wood
Floor: Wood

Below: SWV. For many years railway companies produced sets of permanently coupled paired wagons, thus utilising short wheelbased vehicles for the conveyance of long loads. In BR days these have been confined to a fleet of twin bolsters, formed by permanently coupling pairs of *Lowfit Single Bolster,* or *Conflat A* wagons. The few remaining sets are all vacuum braked and formed mainly of *Lowfit* vehicles. These twin sets retain both vehicle numbers. An unfitted example (now withdrawn) is shown here with former *Lowfit* No B452676 nearest to the camera. *Michael Rhodes*

Group U: Uncovered Open Bulk Carriers

Code	*Description*	*Capacity*	*Brake Type*	*Remarks*
UCV	China clay	13t	Vacuum	Some have hoods
USV	Sand	13t	Vacuum	
UTV	Timber	17t	Vacuum	
UUV	Timber pulp	22.5t	Vacuum	Ex plate
UYV	Anhydrite	25.5t	Vacuum	
UZP	Oxide	25.5t	Vacuum piped	

Below: UCV. Between 1954 and 1960 BR introduced 775 13 tonne five plank china-clay wagons to supplement and replace various ancient GWR vehicles operating in the Cornwall area. Vehicles are intended to transport powdered clay from the quarries to either the docks at Fowey, or industrial users in the south or midlands; normally trains operate in block formations. Although the official livery for these wagons is bauxite, due to the nature of their work the usual colour is chalk white. Nos B743632 and B743329 were photographed at Plymouth Friary yard. *Colin J. Marsden*

Right: UCV. To facilitate off-loading at Fowey Docks wagons have end tippler doors, whereby the wagon can tip its load straight into the hold of a boat from a discharge ramp; standard side flap doors are also provided. Many of the UCV fleet have hoods or tent tops to protect their load from inclement weather and from being blow out of the wagons when in transit. These hoods are normally blue, and when travelling resemble a train of tents on the move! *Clayhood* No B743615 is illustrated here. *Colin J. Marsden*

UCV
CLAY HOOD
B743615

Above: UUV. The standard BR designed *Plate* wagon has often been the subject of conversion and modification work, being used in a number of differing applications. One of the most interesting came in 1965 when a fleet of former *Plate* wagons previously used on the *Condor* freight service, were converted at Barassie works into *Timber-P* vehicles for conveying rough timber logs from Crianlarich to Corpach. During rebuilding work most of the vehicle except the frame was scrapped, and new tall wooden plank ends and a central column were fitted, with side restraining steel straps to prevent the logs from falling out. Over 50 vehicles of this type were converted and all retained their *Plate* number. The livery was bauxite brown with white lettering. No B933233 is shown here when new, but the majority of this type of vehicle have now been withdrawn. *British Railways*

Below: UYV. 150 *Anhydrite* hopper wagons were constructed during 1954 by BR at Shildon for the conveyance of Anhydrite between Newbiggin, near Appleby, and Widnes. Each vehicle can take a 25 tonne load and is usually covered with a light coloured tarpaulin. Fittings include clasp type vacuum brakes and roller bearing wheelsets. In this view we see a train of UYV wagons heading along the S&C line towards Carlisle away from Newbiggin. *Colin J. Marsden*

Group V: Vans

Code	*Description*	*Capacity*	*Brake Type*	*Remarks*
VAB	Van — full length doors, ventilated	20.5-30t	Air, vacuum piped	
VBA	Van — full length doors, non-ventilated	29.5-39t	Air	
VBB	Van — full length doors, non-ventilated	28.5-29.5t	Air, vacuum piped	
VCA	Van — centre doors non-ventilated	29-29.5t	Air	
VCB	Van — centre doors non-ventilated	29-30t	Air, vacuum piped	
VDA	Van — centre/end cupboard doors — non-ventilated	25t	Air	
VDB	Van — centre/end cupboard doors — non-ventilated	25t	Air, vacuum piped	
VEA	Van — roller bearing fitted	12t	Air	Ex vanwide
VEV	Van — Vanwide, roller bearing fitted	12t	Vacuum	
VFW	Fish Van	12t	Vacuum, air piped	
VGA	Van — full length doors, non-ventilated	28t	Air	
VIX	Ferry van/motor car van to continental requirements	14.2-25t	Dual	
VJX	Ferry Van	25t	Dual	Ex VIX
VMV	Van/Vanwide — plain bearings	12t	Vacuum	Some vehicles have a translucent roof
VPB	Palvan	10t	Air, vacuum piped	
VQB	Palvan	22.5t	Air, vacuum piped	
VQV	Palvan	22.5t	Vacuum	
VQW	Palvan	22.5t	Vacuum, air piped	Some vehicles have curtain sides
VVB	Van-box	11.5t	Air, vacuum piped	
VVV	Van-box	12t	Vacuum	
VWV	Vanwide 9ft sliding doors, ventilated	12t	Vacuum	Some vehicles have translucent roof

Above: COV AB (VAB). The first of the modern purpose built air braked van fleet was built by BR at Ashford in 1966, when experimental vehicle No B787479, a 20ft 9in wheelbased, curtain sided palette van emerged, fitted with air brakes and a through vacuum pipe. A production run of over 200 hard sided vans of the same basic design followed. The production fleet was due to be numbered in the B787480-B787699 range, however before the first vehicle emerged the 'new' air braked numbering system was introduced and the vehicles became 200000-200208. Painted in bauxite livery and sporting the code name *COV AB* No 200017 is seen when new. *British Railways*

Below: VBA. After the initial batch of vehicles many subsequent builds were authorised incorporating design improvements. The VB fleet entered traffic during the late 1960s and were the same basic design as the original *COV AB*. The VA and VB types both have two side access doors each giving a 15ft 9in opening. Bodies are constructed of steel and wood and the vehicle weighs 16.6 tonne when in tare condition; capable of carrying up to 29.5 tonne when fully loaded. These vans are especially suitable for palletised loads and have a strengthened floor for pallet truck operation. The photograph shows No 200249 painted in *Railfreight* red/grey livery. *Colin J. Marsden*

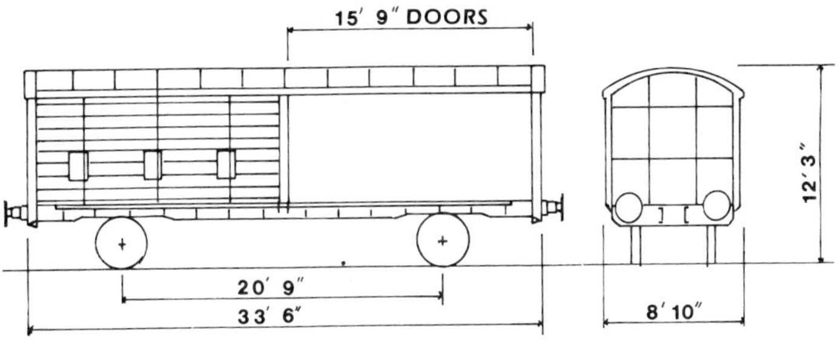

VAB, VBA, VBB
Tare: 16.6 tonne
Carrying capacity: 29.5 tonne at 60mph
24.5 tonne at 75mph
Cubic capacity: 59.4cu m
Minimum curve: 41m
Body: Steel and wood
Floor: Wood

Below: VBB. The VBB variant on the VB wagon type is fitted with a through vacuum pipe and is the only difference to the straight air braked vehicle. This fitting was made to enable the new vans to operate in vacuum train formations which was still the norm. No 200223 in bauxite livery is seen at the Railway Technical Centre Derby where special HSFV wheel sets have been installed for evaluation. The cables around the lamp iron would be connected to instruments in a test car when undergoing road tests. *Colin J. Marsden*

Above: VBB. Although standard livery for the *Railfreight* fleet is now red/grey a considerable number of vehicles are still in traffic painted in bauxite. This will change as and when vehicles pass through works for a classified overhaul. VBB No 200237 clearly shows one of its doors in a fully open position when photographed at Hartlepool during 1982. To open the wagon doors the three recessed handles have to be operated in the correct order. *Colin J. Marsden*

Below: VCA. Although of the same general design as the previous air braked van types, the VCA fleet have a different door arrangement, with two sliding doors giving an opening of 16ft. The majority of these vehicles are painted in the modern *Railfreight* livery. As before, this fleet can carry up to 29 tonne of general goods at a maximum of 60mph, however if a faster speed of 75mph is required the load must be restricted to 24 tonne. VCA No 200517 is seen at Doncaster. *Colin J. Marsden*

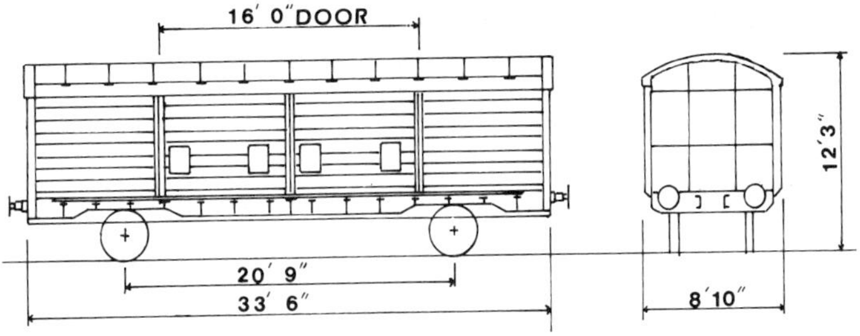

VCA
Tare: 17.5 tonne
Carrying capacity: 29 tonne at 60mph
24 tonne at 75mph
Cubic capacity: 59.4cu m
Minimum curve: 41m
Body: Steel and wood
Floor: Wood

Right: VDA. End detail of VDA type wagon. These are the only examples of the modern air braked van fleet that have outside ribbing on their body ends, and therefore immediately recognisable. The buffer beam layout is very simple and apart from the screw coupling only air brake and main reservoir pipes are present. *Colin J. Marsden*

Above: VDA. A further development of the standard 20ft 9in wheelbase wagon design is this fleet of VDA vans having three door openings on each side, however all cannot be opened at the same time. Both ends have 8ft 1in openings with two hinged doors flanking two centre sliding doors giving a total 16ft 3in opening. This fleet is intended to carry up to 24.5 tonne and weighs 16.25 tonne when in tare condition. Vehicles are recognisable by a deep section to the frame between the bogies. No 210271 is illustrated here.
Colin J. Marsden

Below: VDA. The VDA fleet commenced service from mid-1975 and there are currently over 1,000 of this type in use. Over the years much has been done to try and improve the ride of four-wheeled freight stock and during the late 1960s the Railway Technical Centre at Derby was engaged in producing a high speed freight vehicle (HSFV) suspension system, and during 1976 100 VDA wagons were fitted with HSFV-6 suspension in place of the single long link type, greatly improving the vehicle ride. The new high speed suspension named *Taperlite* is seen here on VDA No 201081. *British Railways*

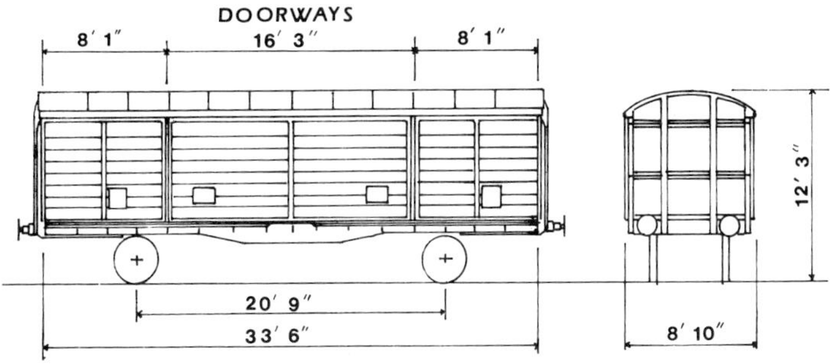

VDA
Tare: 16.25 tonne
Carrying capacity: 24.5 tonne at 75mph
Minimum curve: 41m
Body: Steel and wood
Floor: Wood

Below: VEA, VEV. The long established short wheelbased general goods van, capable of taking a 12 tonne load, has been part of the railway scene for practically the last 100 years. The VE group now has two brake types vacuum and air, with the bodywork for both types basically the same, mounted on a 17ft 6in frame with a 10ft wheelbase. There are two sliding doors giving a 9ft opening. To cater for the modern air brake fitted network a number of old vehicles were rebuilt to modern standards during the late 1970s and fitted for air brake operation, roller bearing axles, repainted in modern *Railfreight* livery and classified VEA. These vehicles are designed to carry general mechandise and are especially intended for use in areas where tight curvature prevails. VEA No 230133 is seen here. *Colin J. Marsden*

VEA, VEV

Above: VEV. The most modern of the standard 10ft wheelbase vans were constructed during 1963. Vehicles fitted with roller bearing wheelsets are classified VEV and are the forerunners to the modern air braked wagon fleet, with wide opening doors suitable for palletised loading, utilising fork lift trucks. All vehicles of this group are ventilated with some having translucent roofs. No B784032 is seen here. *Colin J. Marsden*

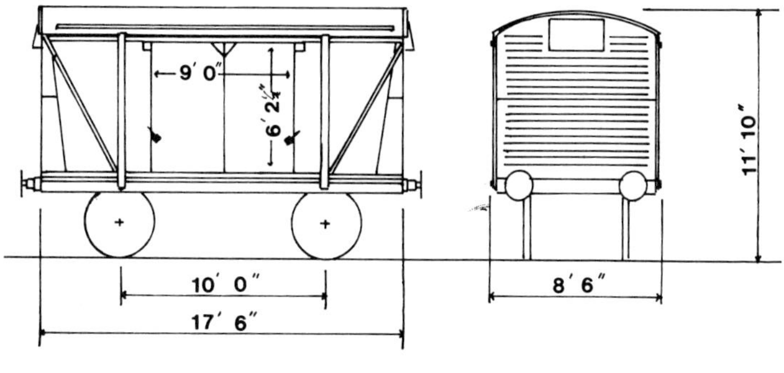

VEA, VEV
Tare: 7.7 tonne
Carrying capacity: 12 tonne
Cubic capacity: 27.7cu m
Minimum curve: 21m
Body: Wood sides with steel ends
Floor: Wood

Above: VGA. The most modern *Railfreight* van is the VGA, having a 29ft 6in wheelbase. These vehicles were designed with the expanding *Speedlink* wagon-load service in mind, being fitted with two 7ft $2\frac{1}{2}$in high and 20ft $8\frac{1}{2}$in long doors on each side, giving half length access at a time. Floor mounted lashing points enable part loads to be secured for safe transit. When loaded to the maximum 29 tonne the permitted speed is 60mph, but if the load is kept to 24 tonne the speed increases to 75mph. These vehicles are in unpainted aluminium with the *Railfreight* or *Speedlink* legend on the sides in white. No 210400 is illustrated here. *Colin J. Marsden*

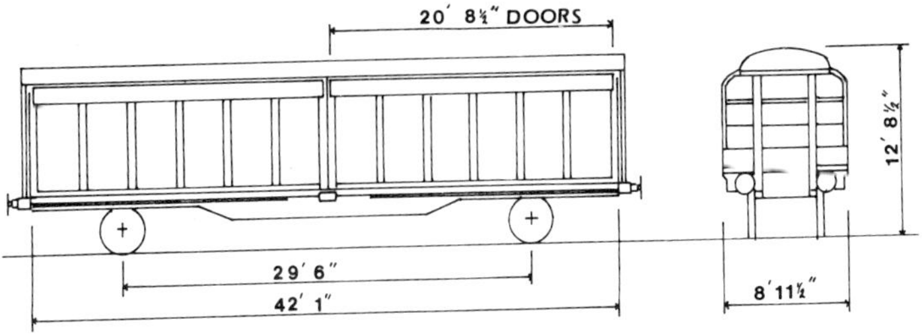

VGA
Tare: 17 tonne
Carrying capacity: 29 tonne at 60mph
24 tonne at 75mph
Minimum curve: 50m
Body: Aluminium and steel
Floor: Wood

VIX, VJX, ILB, ILX

Above: VIX, VJX. The British owned ferry wagons built to UIC standards and designed to operate between Britain and the Continent are classified as VIX. Vehicles have a single large central sliding door opening to 12ft 9in with four ventilation panels on each side. These vans were painted in bauxite livery with a *Through to the Continent* legend on their sides, as illustrated by No 2380 341-4, but in recent years many have appeared with *Railfreight International* insignia. As these vehicles conform to international standards anchor points are provided. A large number of these vehicles are used as Railway Operating Department Stock for barrier use. During 1983 a number of these vehicles were taken out of the international fleet, for use in England as normal 25 tonne capacity vans, and given their BR B prefix numbers in the B787xxx range. TOPS coding has been amended to VJX. *Colin J. Marsden*

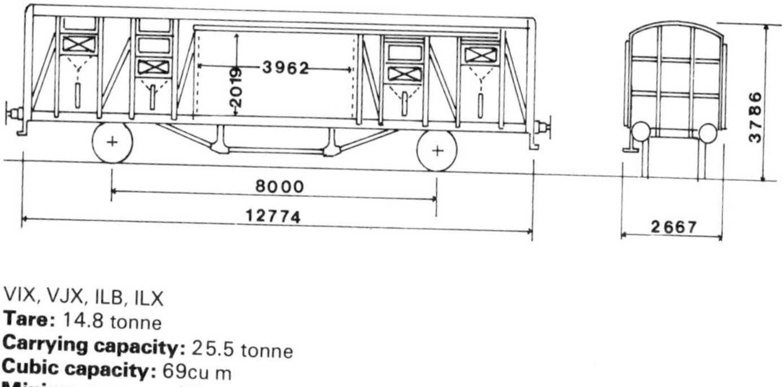

VIX, VJX, ILB, ILX
Tare: 14.8 tonne
Carrying capacity: 25.5 tonne
Cubic capacity: 69cu m
Minimum curve: 46m
Body: Wood and steel
Floor: Wood

Above: VMV. Looking similar to the previously illustrated VEV vehicle, we now take a look at the VMV group. These are identical to the VE type except for carrying conventional oil axle boxes. All VMVs are fitted with clasp type brakes, and can convey a load of 12 tonne. Different door arrangements may be found on some examples. The majority of the remaining vehicles in this group are from the 1960 builds and scheduled for early withdrawal. No B783444 was photographed at Plymouth Friary. *Colin J. Marsden*

Below: VQB, VQV, VQW. The TOPS code VQ is allocated to the *Palvan* type of van vehicle. These are long wheelbase wagons designed for the transportation of palletised goods, normally operating between factory sites and often used within the car industry. The *Palvan* in the illustration is being loaded by forklift truck and bears the legend — 'Dagenham-Halewood — daily by rail', indicating the train is in use by the Ford Motor Company for the transportation of car parts; each vehicle can carry 22 tonne of cargo. *British Railways*

VVV, VWV

Above: VVV. The standard 12 ton box van, already seen in the VE and VM groups can also fall into the TOPS group VVV. This code is given to ordinary vans fitted with push brakes and hinged doors in place of the sliding type on VE and VM vehicles. The standard box van is formed of plywood sheets fitted to a steel frame with steel reinforced ends to prevent damage from any shifting load. VVV No B775947 built during 1957 — one of 700 from lot No 3007 — was photographed at York. *Colin J. Marsden*

Below: VWV. The final TOPS code given to the 12 tonne box van is VWV, used for *Vanwides* with 9ft sliding doors and ventilated. Again this group is mainly formed of vehicles manufactured during the early 1960s and fitted with clasp type brakes. A number of vehicles within the group have translucent roofs, fitted to improve working conditions inside the van during loading and unloading. Vehicles so fitted are normally labelled so on the outside, as shown here on No B784837 at Guildford. This van also bears the legend 'Return to Avonmouth Docks W, for R. Silcock and Sons Ltd'. *Colin J. Marsden*

Group X: Exceptional Vehicles and Special Purpose Vehicles

Code	*Description*	*Capacity*	*Brake Type*	*Remarks*
XAO	Armour plate 'EL' 'ET' 'WE'	40-56t	Unfitted	
XAP	Armour plate 'AB' 'ET'	40-56t	Vacuum piped	
XAQ	Armour plate 'ET'	56t	Air piped	
XCO	Concrete beam	32.5-51t	Unfitted	Ex armour plate, flat, gunset, transformer, trestle or weltrol vehicle
XCP	Concrete beam	51t	Vacuum piped	Ex flat
XDV	Flat 'ED'	12t	Vacuum	
XFO	Flat 'ET'/Warflat	46-51t	Unfitted	
XFQ	Flat 'ET'	46t	Air piped	
XFV	Flat 'WC'/Atomic flask	51t	Vacuum	
XIQ	Lowmac EU (continental standard)	25t	Air piped	
XIR	Rectank (continental standard)	38.5t	Dual piped	
XIW	Flatrol/Weltrol (continental standard)	21.3-55.5t	Vacuum, air piped	
XJB	Atomic Flask Carrier	77.5t	Air, vacuum piped	
XKB	Flatrol (for nuclear flask)	56t	Air, vacuum piped	
XKV	Flatrol (for nuclear flask)	54t	Vacuum	
XLO	Lowmac 'MR'	21.5t	Unfitted	
XLP	Lowmac 'AB'	21.5t	Vacuum piped	
XLV	Lowmac EO, EP, EQ, ES, ET, MK, MO, MU, SG, WB, WV, WW	20.5-25.2t	Vacuum	
XLW	Lowmac EO, EQ, ER, ET, MK, MU, SC, SG, WW	22.5-25.5t	Vacuum, air piped	
XMA	Car Transporter — Cartic	8t	Air	
XNB	Flask carrier — 2 axle	34t	Air, vacuum piped	
XPO	Flatrol, ELL, ET, MHH, WX	40.5-81.5t	Unfitted	
XPP	Flatrol MCC, MDD	40.5t	Vacuum piped	
XRP	Rectank ED, MD, WC	38.5t	Vacuum piped	
XSO	Concrete Beam Spacer	—	Unfitted	Ex glass MD
XSP	Concrete Beam Spacer	—	Vacuum piped	Ex flatrol
XTP	Trestrol AO, EO	40.5t	Vacuum piped	
XVA	Trestrol plate 'AB', 'ED'	43t	Air	
XVV	Trestrol plate 'AB', 'ED'	43t	Vacuum	
XWO	Weltrol MC	20.5t	Unfitted	
XWP	Weltrol EL, MC, MV, MX, WP, WW, WGG	30.5-40.5t	Vacuum piped	
XWQ	Weltrol WBB	25.5t	Air piped	
XXO	Twin girder WG	20.5t	Unfitted	
XYP	Boiler/Transformer MA	61-76t	Vacuum piped	

XKB

XYV	Boiler	76t	Vacuum
XZO	Weltrol/EN/ Transformer MC	110-135t	Unfitted
XZQ	Flat EU/Weltrol	100-120t	Air piped

As previously mentioned, the TOPS usage of the letter 'X' in wagon descriptions usually refers to an exceptional vehicle, or one used for a highly specialised purpose. The majority of the vehicles are one off examples so due to restricted space it has been decided to include a small selection of vehicles that hold the most interest. A number of bogie and flat wagons of this group are being handed over to the departmental fleet, and a fuller study of their type and application will be given in *Rolling Stock Recognition* covering departmental stock. Certain vehicles in this group are used for transportation of classified items and descriptions of vehicles concerned, their identities, owners and numbers cannot therefore be included.

Above: XKB. Some of the most interesting wagons classified under the X group, both to the modeller and general enthusiast are those that little information can be disclosed about. Under the TOPS system the code XKB is described as *Flatrol MJ.* There were 24 of these vehicles built between 1960-63, numbered in the B900509-532 range, these were well wagons with a special central assembly for taking a nuclear flask and when introduced these vehicles were mounted on six-wheel bogies. The vehicles were rebuilt at BREL Shildon during 1981 and fitted with Y25 four-wheeled bogies, a new superstructure and air brake equipment. In their new guise they could carry 56 tonne. No B900528 on a nuclear train at Sellafield is illustrated. *Colin J. Marsden*

Below: In 1969 Shildon wagon works produced six flasks to lot No 3697, classified as *Flatrol MJJ.* These vehicles were numbered B550000-B550005 and were well wagons carried on heavy cast framed four-wheeled bogies, and fitted for air brake operation with a through vacuum pipe. No B550004 is illustrated. *Colin J. Marsden*

Above: XRP. TOPS code XRP is for *Rectank ED, MD, WC* vehicles, all are of the same basic design and carry a maximum load of 38.5 tonne. The vehicles owe their design to War Department stock, introduced during World War 2 for transporting Army tanks. The present fleet of vehicles was introduced during 1960 and built by BR at Swindon. Today they have a much wider usage and in addition to Army equipment can convey railway or engineering plant. A distinguishing feature of both this and the *Lowmac* type (classified as XL) is their sloping ends as a drive-on, drive-off facility, unlike the *Flatrol* vehicles which have a deep central well. *Rectank ED* No B909053 is shown. *Colin J. Marsden*

Below: XVA, XVV. From 1981 a fleet of 51 *Trestle Plate EDs* were constructed by BREL to augment the ageing vacuum braked examples. The purpose of these vehicles is to convey steel plate and other flat goods of more than 8ft width. The flat load is carried at 48° against an 8ft 6in high skeleton frame; these vehicles are 52ft long with a 40ft wheelbase, and able to carry up to 43 tonne. The load is secured by hand operated tensioners. No 990010 is shown here loaded with steel. XVV type not illustrated. These vehicles are basically of the same design but can only carry a 42.5 tonne load. *Michael J. Collins*

XKB, XVA

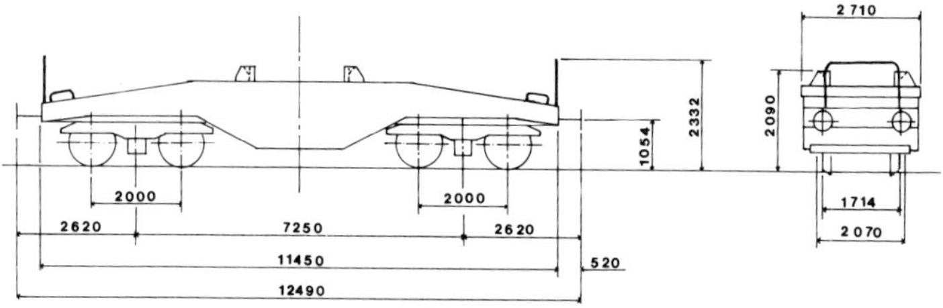

XKB
Tare: 23 tonne
Carrying capacity: 56 tonne
Minimum curve: 31m
Body: Steel
Floor: Steel and wood

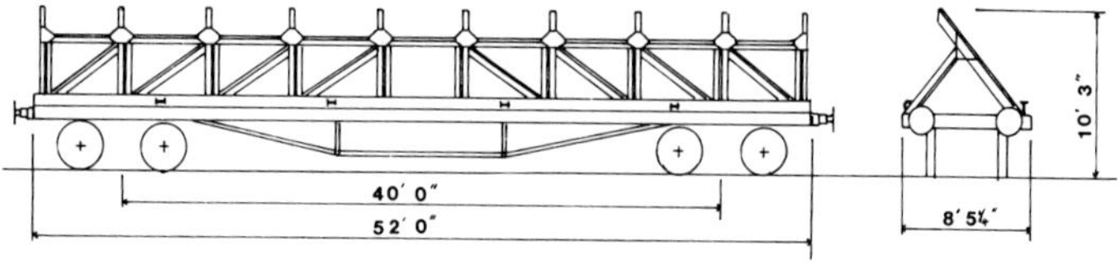

XVA
Tare: 22 tonne
Carrying capacity: 43 tonne
Minimum curve: 22m
Body: Steel
Floor: Wood

Group P: Privately Owned Wagons — Excluding Tanks

Code	*Description*	*Capacity*	*Brake Type*	*Remarks*
PAA	Covered hopper/china clay/grain/lime/ dolomite — 2 axle	31.5-37.5t	Air	
PAB	Covered hopper — 2 axle	32-32.5T	Air, vacuum piped	
PAF	Grain — 2 axle	23.5-24.5t	Vacuum (AFI)	
PAO	Covered hopper/grain — 2 axle	20.5-32.5t	Unfitted	
PBA	Covered hopper—bogie	50.5-71t	Air	
PCA	Presflo — 2 axle	32-38.5t	Air	
PCB	Presflo — 2 axle	33-34.5t	Air, vacuum piped	
PCF	Presflo — 2 axle	34.5t	Vacuum (AFI)	
PCV	Presflo — 2 axle	22-27.5t	Vacuum	
PDA	Presflo — bogie	79t	Air	
PEA	Bulk Tippler	26t	Air	
PEB	Bulk Tippler	26t	Air, vacuum piped	
PFA	Flat (Special) 2 axle	29-39t	Air	
PFB	Flat (Special) Bogie/ carflat/tank carrier/ trolley well (Weltrol)	10-51t	Air, vacuum piped	
PFO	Trolley Flat (Flatrol)	41t	Unfitted	
PFV	Flat/tank carrying bogie /trolley well (Weltrol)	50-51t	Vacuum	
PFW	Carflat — bogie	12t	Vacuum, air piped	
PFX	Flat (Special) bogie	54t	Dual	
PGA	Aggregate/limestone/ salt/iron pellets/ gypsum — hopper — 2 axle	32-39t	Air	
PGB	Salt hopper — 2 axle	32.5-33.5t	Air, vacuum piped	
PGO	Sand hopper — 2 axle	27.3t	Unfitted	
PHA	Aggregate/ballast hopper — bogie	28-75t	Air	
PHO	Ironstone hopper — bogie	68t	Unfitted	
PHV	Limestone hopper — bogie	48t	Vacuum	
PIA	Large van (continental standard)/well wagon	54-101t	Air	
PIB	Hooded Coil/covered hopper	49-52t	Air, vacuum piped	
PJA	Car transporter — Cartic	8t	Air	Inner and outer vehicles
PJB	Car transporter — Cartic	8t	Air, vacuum piped	Inner and outer vehicles
PKA	Artic 3 axle car transporter, two tier	9.5-13.5t	Air	
PLA	2 tier car carrier	12t	Air	
PMO	Mineral	16.5-22.5t	Unfitted	Some have doors while others do not

Code	*Description*	*Capacity*	*Brake Type*	*Remarks*
PNA	Open pallet	40t	Air	
POA	Open scrap	37t	Air	
POO	Soda ash	16.5-20t	Unfitted	
POV	High	13t	Vacuum	
PPV	Escort coach — bogie/ weedkilling train coach — bogie	2-15t	Vacuum	
PSA	Ironstone tippler — 2 axle	34.5t	Air	
PSO	Ironstone tippler — 2 axle	20-27.5t	Unfitted	
PTA	Ironstone tippler — bogie	77.5t	Air	
PVB	Palvan — 2 axle	24.5-36.5t	Air, vacuum piped	Some vehicles have curtain sides
PVF	Palvan	26t	Vacuum (AFI)	
PVV	Van	12-22t	Vacuum	
PWA	Palvan — bogie (with curtain sides)	49-57t	Air	
PXA	Match/bogie steel	10-77t	Air	
PXB	Flask wagon — 2 axle	46t	Air, vacuum piped	
PXF	Tube	25t	Vacuum (AFI)	
PXO	Bogie bolster C/Ingot, bogie — with insulated ends and sides	30.5-71t	Unfitted	
PXP	Ramp vehicle	11t	Vacuum piped	
PXQ	Match vehicle	5-15t	Air piped	
PXV	Hot core — road and railer/coil carrier 2 axle/tube/single bolster/mobile training vehicle	13-96.5t	Vacuum	
PXW	Ramp vehicle	10t	Air, vacuum piped	

In recent years there has been a large rise in the number of privately owned wagons operating on the BR network. The majority operate in block train formations with vehicles either being owned by the operating company or by one of the wagon leasing firms. To carry privately owned fleets on the BR TOPS computer system, each vehicle is given a three letter TOPS code in the same way as BR's own wagon fleet. However the degree of information given by the code is less, for example, all privately owned covered hopper wagons are classified as PA, whether they carry grain, stone or lime, and can weigh anything between 20 and 37 tonne. Information about individual vehicles is sometimes difficult to obtain as many owners are not prepared to give such information. However, a broad outline of the vast privately owned fleets operating on the BR network is included with a brief description of vehicles used and livery if known.

Above and below: PAA. Group PA is for privately owned covered hopper wagons capable of carrying any load, braking systems of types A, B, F and O are currently in use. In the upper plate a PAA owned by Tiger Rail Leasings and operating for Tullis Russell, a paper manufacturing firm, is shown. This vehicle can carry 37 tonne and is loaded by a roof mounted centre weighted sliding door and discharged by two floor hoppers. The lower plate shows a 37.7 tonne hopper built by W. H. Davis of Langwith Junction and owned by British Industrial Sand. This vehicle has a pneumatically operated roof and hopper doors and is used for transporting both wet and dry sand. A total of 40 wagons of this type is now in traffic and normally operate on East Anglian-South Yorkshire duties in block formations. *Colin J. Marsden/Michael J. Collins*

Above and below: PAF. A number of privately owned vehicles carry the letters BRT, these are the initials for British Railway Traffic & Electric Company, a wagon leasing company. These PAF vehicles are often referred to as whisky wagons, but in fact only convey malt barley grain from which the whisky is distilled. All vehicles have a central opening roof section for loading, with discharge through doors in the floor. Various liveries can be found on these vehicles including blue and brown and some carry advertisements on their sides, but this should not be taken as an indication that the vehicle is operating for that company as they are purely advertisements. In the upper view BRT No 5816 in original condition is illustrated, while the lower plate shows No BRT 7586 devoid of its TOPS code.
British Railways/Barry J. Nicolle

Above: PBA. Group PB is for large bogie covered hopper wagons. The vehicle illustrated is owned by Tiger Rail Leasings and operating for English China Clay (ECC) for the transport of china clay from Cornwall to various railheads throughout the country. These vehicles started to emerge during 1982 being manufactured in France by Etablissements Fauvet-Girel and numbered in the TRL11600-TRL11634 range, and will replace the BR operated 12.5 tonne clay wagons previously working in Cornwall. These vehicles have a large contre balanced roof section for easy loading. The first wagon of the build No TRL11600 is shown here at Hartlepool.
Colin J. Marsden

Right: Side information panel and owner's address on side of PBA wagon. *Colin J. Marsden*

Above: PCA. This group is for Pressure flow or *Presflo* hoppers mounted on a two-axle chassis. Two distinct body designs exist, those having a dropped centre and those with straight tanks used to transport powdered cement. This view of PCA No 10913 shows the dropped centre type owned by Blue Circle Cement, thus the prefix BCC to the vehicle number. This wagon was built by Procor Ltd and can carry up to 37.8 tonne.
Colin J. Marsden

Below: PCA. This PCA introduced during 1981-82 for use by the Rugby Cement Company is built and owned by Procor Ltd and on long term lease to Rugby Cement. The pressure discharge equipment is visible between the wheels. This wagon, No PR10040 can carry up to 36.8 tonne and is normally to be found operating in block train formations, although local wagon traffic does sometimes generate from Halling Cement Works in Kent. 25 vehicles of this type Nos PR10025-10049 are in traffic. *Colin J. Marsden*

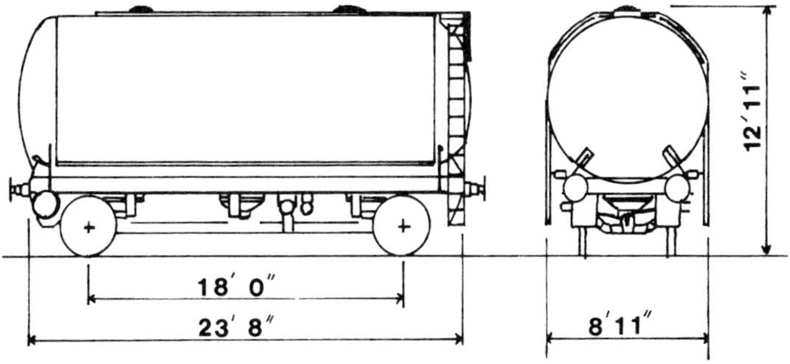

PCA
Tare: 14 tonne
Carrying capacity: 32-38.5 tonne
Cubic capacity: 36cu m (approx)
Minimum curve: 80m
Body: Steel

Below: PDA. To facilitate the transportation of larger loads of powdered cement a fleet of bogie cement hoppers, formed of two dropped centre type containers (the PC type) mounted on a bogie underframe, were built by Metro-Cammell in 1969. These large 79 tonne capacity wagons are also of the pressure discharge type and usually to be seen operating in block formations with the four-wheeled type. BCC No 9727 is seen here.
Colin J. Marsden

Above: PFA. All privately owned flat wagons are classified as PF, this covers a weight range of between 10 and 51 tonne. Various brake types of Groups A, B, O, V and W exist. PFA No F93000 is illustrated here; this vehicle is owned by Freightliners and is a prototype two-axle 40ft flat specially developed for the growing number of 40ft containers. It can take a 30 tonne load and weighs only 11 tonne when in tare condition. *Colin J. Marsden*

Below: PFB. This vehicle classified as PFB is a carflat owned by Rail Leasing (RLS) and is a former locomotive hauled Mk 1 coach underframe converted to transport motor cars. It is fitted with air brakes and through vacuum pipe, painted in yellow and usually to be found on the Sheerness branch. The vehicle carries number RLS 94407 and is seen here at Queenborough. *Colin J. Marsden*

Above and below: The Ministry of Defence operate a sizeable fleet of railway wagons for transporting military equipment, these are usually painted green and carry the MODA prefix to their numbers. In the upper illustration we see MODA 95531 a 54 tonne bogie well wagon fitted with air brakes and through vacuum pipe. Note the screw jacks at the end of this vehicle under the buffer, which are screwed down to rail height when end loading/ unloading takes place. The lower illustration shows a train of MOD vehicles of various types. The first vehicle is a runner and is of special interest. It is both air and vacuum piped, has screw loading/unloading jacks and hinged buffers enabling the vehicle to be placed hard against a loading dock. Note the white painted spanner on the buffer beam for releasing the buffers. This vehicle carries No MODA 95000. *Both: Colin J. Marsden*

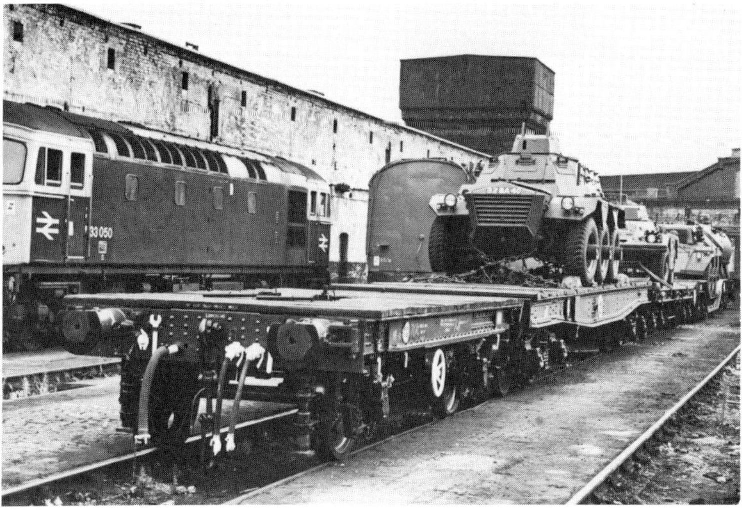

PGA

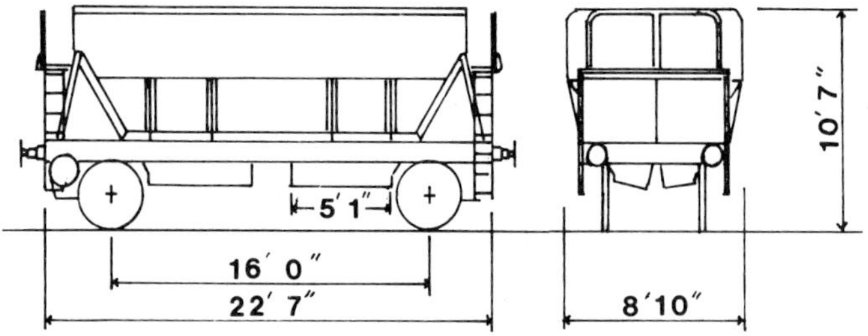

PGA
Tare: 13 tonne (varies on design)
Carrying capacity: 34-38 tonne (varies on design)
Cubic capacity: 36cu m
Minimum curve: 50m
Body: Steel

Below: PGA. Perhaps the most numerous of private owner vehicles is the hopper, fitted on a two-axle chassis; the majority have air brake equipment but a handful are of brake type B and O. It is usual for the PGA hopper fleet to operate in block formations and usually carry the operator's name on the side. *Tilcon* PGA No TCS 14400 is shown here when brand new. *British Railways*

Above: PGA. The purpose of the hopper wagon is that it can be loaded by overhead conveyor or grab and discharged via floor mounted doors. This is normally carried out over a stage where the load is dropped down a chute on to a conveyor belt and away for storage or dispatch. PGA No REDA 14816 owned and operated by Redland Roadstone, is seen on the discharge facility at Kennett Redland Roadstone terminal. This example can carry up to 38 tonne of stone. *Barry J. Nicolle*

Above: PGA. The PGA type hopper is operated by a vast number of companies, some wagons are owned by their operators whilst others are owned by Procor Wagon Leasings. The owner can be determined by reference to the number prefix; if it is made up of letters from the operator's name it will be owned by them, ie: Tilbury — TBR, Redland — REDA, but if the prefix is PR it is owned by Procor Wagon Leasings. Various designs of wagon exist within the PGA group some having higher sides than others, different discharge arrangements and end steps. Tilbury Roadstone Ltd owned and operated No TBR 14501 is illustrated here, this wagon was constructed by Procor for Tilbury and can carry 36.9 tonne. *British Railways*

Below: PGA. What can only be described as a smaller version of the HAA hopper used on the MGR circuits is this fleet of PGA hoppers used for the conveyance of Gypsum from Mountfield to Northfleet; the vehicles operate under the BCC private owner code and are normally found in block formations. The vehicles have roller bearing axles, air brake equipment and carry 34 tonne of gypsum, automatic discharge equipment being fitted. A feature of these wagons is the painted number on the unpainted body side in black. *Michael J. Collins*

Above: PHA. Some of the largest bogie hoppers in use is a fleet of French built vehicles operating for Francis Parker between Lavant and Drayton on the SR. These vehicles have the unusual facility of side discharge and are therefore heavily restricted as to where they may operate. Francis Parker bogie hopper No 1 carrying TOPS code and number PHA FR17001 is shown here; the vehicle can carry 68.5 tonne and when loaded weighs 91.5 tonne. *Les Bertram*

Below: PIA. The largest British privately owned vehicles in operation are a small fleet of 101 tonne well wagons owned by British Nuclear Fuels and used for the transportation of large and heavy equipment to and from the Windscale plant in Cumbria. Vehicles are mounted on eight-wheel articulated bogies at each end and are manufactured to international standards, enabling them to operate through to the Continent. Our picture shows three of these vehicles with No 21 70 0998 002-0 leading. *Colin J. Marsden*

Above: PIA. The group PIA also covers large bogie vans owned by private British operators for continental use. These high capacity ferry wagons were designed to meet the demands of developing rail freight business between the UK and the Continent and have a carrying capacity of 54 tonne. Three sliding door panels on each side provide a plain surface when closed, but in the open position give access to one-third of the vehicle at a time. These vehicles are intended to convey general merchandise and palletised traffic. The vehicle in the illustration is operated by Danzas. *British Railways*

Below: PIB. A fleet of 52.5 tonne bogie hooded ferry wagons are operated by the VTG Company of London. Built to UIC standards and designed for through traffic to the Continent, they are intended for the cartage of coils and other heavy goods. Three sliding hoods enable up to two-thirds of the vehicle to be opened for loading/unloading at one time. The wagon floor has five wells to accommodate coils, each with a fold over flap, which provides support for the cargo when open and a flat floor when closed. Air brakes and a through vacuum pipe are provided. A VTG hooded ferry wagon is seen here behind a Class 86 at Willesden. *Colin J. Marsden*

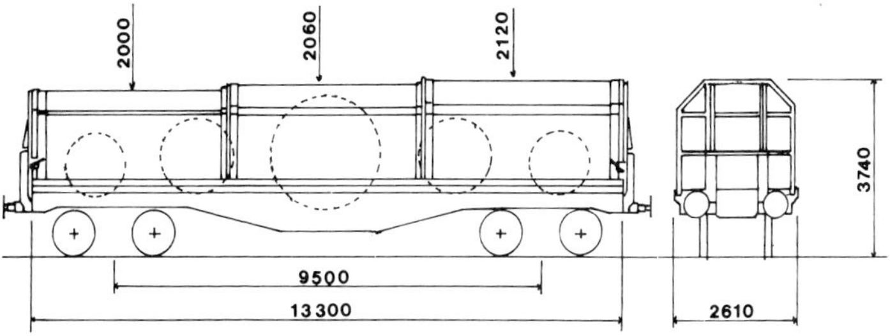

PIA, PIB
Tare: 27.5 tonne
Carrying capacity: 52.5 tonne
Floor area: 27.5sq m
Minimum curve: 35m
Body: Steel/aluminium
Floor: Wood

Below: PJA/PJB. Privately owned *Cartic* 4 sets are classified as PJA, whilst BR owned sets are classified as XMA. These vehicles are permanently coupled in four-car sets having articulated bogies within sets. The *Cartic* 4 is a double deck unit capable of loading most designs of modern motor car. Seven sets of four-car cartics form the average train with a capacity of 170 cars. The PJA set shown here is owned by Silcock & Colling and is at Crewe loaded with new Ford Escort cars. PJB/XMA types are of the same design but not illustrated. *Barry J. Nicolle*

PKA

Above: PKA. A fleet of articulated three-axle two-tier car transporters was introduced during the early 1980s for the transportation of imported cars. The sets have end loading facilities and carry the insignia 'Cartrain' on their sides. No RLS 92029 shown here is owned by Rail Leasings and painted in yellow ochre; photographed at Queenborough. *Colin J. Marsden*

Below: PKA. Basically designed for the transportation of lorries or tall vehicles unable to be loaded on two-tier carriers, is this 26 tonne (total) carrying capacity three-axle articulated set, fitted with clasp air brakes. The vehicles are operated by Rail Leasings and numbered in the range RLS 9230X. Cars RLS92306-7 are seen here painted in yellow livery, at the Railway Technical Centre, Derby during 1982, when tests on the vehicle's design were taking place. *Colin J. Marsden*

Above: PLA. One of the longest freight vehicles operating in the private owner sector, is a fleet of *Procor 80* car transporter vehicles. These two-deck transporters carried on two four-wheel bogies, are designed for block train or speedlink operation, being fully compatible with all other modern air braked rolling stock. Up to 10 average sized family motor cars can be carried on each *Procor 80*. *British Railways*

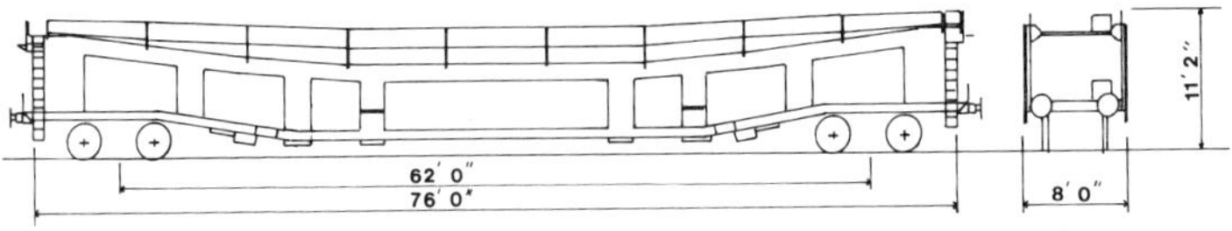

PLA, XMA
Tare: 27 tonne
Carrying capacity: 12 tonne
Minimum curve: 70m
Body: Steel

Above, below and above right: POA. The next three illustrations all show the large scrap steel wagon type, classified PO under the TOPS system, all built and owned by Procor Ltd and on long lease to Sheerness Steel Ltd in whose livery the vehicles are painted. This company is one of the few remaining private steel producers of any size in the country today. The first two plates show wagons constructed in 1975 capable of carrying a 69.4 tonne load and mounted on Gloucester clasp braked bogies. The plate above right shows the latest type to emerge from Procor Ltd for use by the Sheerness Steel Company, this vehicle having modern disc braking and Schlieren primary suspension is of a lighter construction but can carry up to 73.1 tonne. Some bogie flat wagons are also operated by this company. *All Colin J. Marsden*

Below: PSA. Group PS is for *Ironstone* tippler wagons with a carrying weight of between 20 and 38.5 tonne. The vehicle illustrated here owned by Rail Leasings, is a side tippler identifiable by the side hinged doors, and is the first of a new design introduced during 1983 with clasp type air brakes which operate on the single pipe system. Photographed at Eastleigh Open Day carrying number RLS 5400. *Colin J. Marsden*

Above and below: PTA. For the transportion of iron ore to steelworks in South Wales and Scotland, a number of block formations of PTA wagons are in service. These are formed of bogie ironstone tippler wagons manufactured by Redpath Dorman Long, a subsidiary of BSC, and capable of carrying up to 77.5 tonne. All vehicles are mounted on cast steel bogies manufactured by the BSC at their River Don works in Sheffield. There are both inner and outer vehicles; inner vehicles have buck-eye couplings and outer vehicles have a conventional coupling hook for connection to standard BR equipment. In the upper plate is one of the Scottish operated sets with vehicle No BRSV 26686 nearest the camera, while in the lower plate is a South Wales set used between the Port Talbot docks and Llanwern BSC works, discharging its load at Llanwern. *John Tuffs/British Railways*

Above: Close-up of the buckeye coupling arrangement between PTA vehicles. The left hand set is a rotary type whilst the right hand set is of the fixed type, having a white painted uncoupling handle nearest the camera. *British Railways*

Below: For discharge at Llanwern each vehicle is carefully positioned in the rotary tippler unit, it is then completely inverted emptying its cargo into a floor hopper for movement by conveyor. *British Railways*

Above: PVA. A new design of Procor *Palvan* emerged during 1982. The illustration shows the demonstration vehicle numbered PR 6915, it is a 31.6 tonne capacity vehicle with a tare weight of 14.4 tonne, and mounted on a 9m underframe fitted with Bruninghaus and BR long-link suspension. Brake pressure is calculated by a load proportional valve thus reducing wear on all movable components. *Colin J. Marsden*

Below: PVB. Based on the BR *Railfreight* and *Speedlink* van types, Procor Ltd introduced a fleet of air braked, vacuum piped vans during 1981 having four hinged doors on each side which, when open, give access to ⅔ of the vehicle length at one time. Each vehicle weighs 15 tonne when in tare condition and can carry a 35.8 tonne cargo. They are normally on long term lease from Procor and often carry the operator's name on the side. PVB No PR6400 shown here carries the name Benchairs (Western) Ltd when photographed during 1981. *Procor Ltd*

Above and below: PWA. Privately owned *Palvans* usually used for transporting fertilisers are classified PW, all being fitted with air brakes. Two distinct types exist, those fitted with curtain sides and those with eight hinged doors on each side. All vehicles are mounted on Gloucester style bogies and usually operate under the Shellstar or UKF Fertilisers name. A curtain sided vehicle is shown in the upper plate, with a more modern door sided vehicle in the lower illustration. *British Railways/Colin J. Marsden*

Above: PWA. View showing PWA wagon with heavy duty curtain sides in the drawn back position. It will be noted that the vehicle has two loading levels for ease of stacking the fertiliser bags. The vehicle illustrated shows the Shellstar livery and was built by the Gloucester Railway Carriage & Wagon Company and is capable of taking a 56 tonne load. *British Railways*

Below: PXA. Again the X group within the private owner section is for exceptional or special vehicles. For the protection of locomotives from highly dangerous chemicals or goods, barrier vehicles are used. These are normally supplied by BR and operate under TOPS Group R, however there are a small number of privately owned barrier vehicles in existence. This barrier classified PXA is owned by Rail Leasings, a modern air braked steel vehicle weighing 14 tonne having large solid plate ends and normally used between trains of 99.5% neat acid and BR locomotives, operating in the ICI chemicals complex at Seal Sands. This barrier is understood to seldom operate on the BR main line although authorised to do so, and carried No RLS 4916 when photographed during early 1982. *Colin J. Marsden*

Above: PXV. Although privately owned tank wagons are officially classified under the TOPS code T, this vehicle conforming to the general design of a tank wagon is in fact a mobile training vehicle for staff involved in Liquid Petroleum Gas (LPG), not authorised to carry a liquid cargo, and being of a special nature the code is PX. The vehicle is owned by the Shell Company and operates with the number SUKO 50000, having a tare weight of just 18 tonne. It was photographed at Toton depot during the summer of 1983. *Colin J. Marsden*

Below: PXV. This mammoth articulated vehicle operated by the Ministry of Defence conveys material of a classified nature between Ministry of Defence establishments. The vehicle is mounted on two special 12-wheel bogies and can carry in excess of 100 tonne; no further details of this vehicle are available. The train in the illustration was photographed passing Plymouth station during August 1978. *Les Bertram*

Group T: Tank Wagons — Privately Owned

Code	*Description*	*GLW Range*	*Brake Type*	*Remarks*
TBB	Tank bogie	70-79t	Vacuum, air piped	
TBV	Tank bogie	70-79t	Vacuum	
TCA	Tank bogie	80-89t	Air	
TCB	Tank bogie	80-89t	Air, vacuum piped	
TCW	Tank bogie	80-89t	Air, vacuum piped	
TDA	Tank bogie	90-99t	Air	
TDB	Tank bogie	90-99t	Vacuum, air piped	
TEA	Tank bogie	100-+t	Air	
TEB	Tank bogie	100-+t	Air, vacuum piped	
TIA	Tank 2 axle/bogie tank International standard	30-80t	Air	
TIB	Tank 2 axle/bogie tank International standard	30-80t	Air, vacuum piped	
TIG	Tank 2 axle	40-49t	Vacuum (AFI) air piped	
TIQ	Tank 2 axle	30-39t	Air piped	
TIR	Tank 2 axle	30-39t	Dual piped	
TIW	Tank 2 axle	30-49t	Air, vacuum piped	
TIX	Tank 2 axle	30-49t	Dual	
TMV	Tank 3 axle	20-29t	Vacuum	
TRO	Tank 2 axle	20-29t	Unfitted	
TRQ	Tank 2 axle	20-29t	Air piped	
TRV	Tank 2 axle	20-29t	Vacuum	
TSB	Tank 2 axle	30-39t	Vacuum, air piped	
TSF	Tank 2 axle	30-39t	Vacuum (AFI)	
TSO	Tank 2 axle	30-39t	Unfitted	
TSR	Tank 2 axle	30-39t	Dual piped	
TSV	Tank 2 axle	30-39t	Vacuum	
TTA	Tank 2 axle	40-49t	Air	
TTB	Tank 2 axle	40-49t	Air, vacuum piped	
TTF	Tank 2 axle	40-49t	Vacuum (AFI)	
TTG	Tank 2 axle	40-49t	Vacuum (AFI), air piped	
TTV	Tank 2 axle	40-49t	Vacuum	
TTW	Tank 2 axle	40-49t	Air, vacuum piped	
TUA	Tank 2 axle	50-+t	Air	
TUB	Tank 2 axle	50-+t	Vacuum, air piped	

TOPS Group T is for privately owned tank (liquid) carriers. As with privately owned general freight vehicles of Group P, one TOPS code can apply for a number of different vehicle designs and types.

Above: TBV. Code TB is applied to bogie tank vehicles with a gross laden weight (GLW) of up to 79 tonne. This former Murgatroyds tank No T209 now part of the BP fleet, falls into this category with a GLW of 63.2 tonne. It is a vehicle built on frames of sizeable proportions yet has only a short tank. This fleet of tanks, now thought to number about 10, usually operate on the Sandbach-Spondon route. No BPCM 77009 is seen at Spondon.
Michael J. Collins

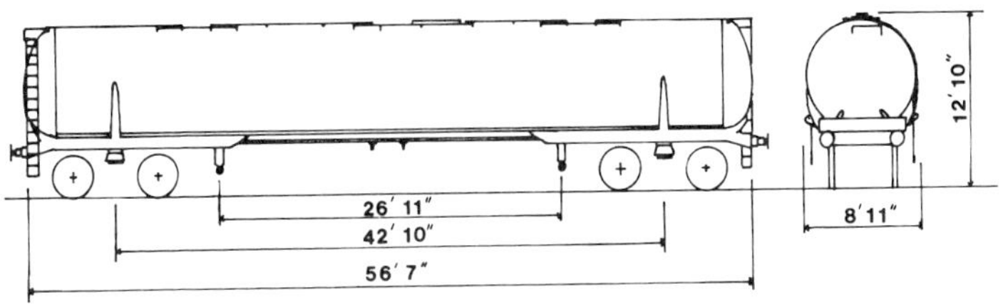

TBB
Tare: 27 tonne
Carrying capacity: 74.6 tonne
Gross capacity: 94,766 litres
Minimum curve: 70m
Body: Steel

Above: TBB. An interesting fleet of vehicles was constructed during 1970 by the Standard Wagon Company of Heywood, Lancashire, for the conveyance of liquid chlorine. Vehicles were mounted on a bogie chassis with a 17ft $1\frac{1}{4}$in wheelbase, able to carry a 40 tonne load with a weight of just over 26 tonne in tare condition; the total length is 34ft 3in. No T240 is illustrated here when new. *Colin J. Marsden Collection*

Below: TEA, TEB. The most common bogie tank type to be observed is the 100 tonne GLW vehicle conforming to the TEA or TEB code. There are a large number of these in operation with various oil and wagon leasing companies all to the same basic design but with detail differences, such as ladder and undertank mounted arrangements. Most of these vehicles are mounted on Gloucester type bogies. TEB No BP83592 illustrated here, is capable of carrying up to 72.5 tonne of liquid cargo at speeds of up to 60mph. *Colin J. Marsden*

Above and below: TIA. Group TI under the TOPS wagon system is for British privately operated tank wagons of all descriptions conforming to the UIC/RIV standard, making them compatible to operate on foreign railways, a number of different brake types of groups A, B, G, Q, R, W and X exist within this type. In the upper illustration: a small four-wheeled tank owned by ICI capable of carrying up to 18.3 tonne of chemical and classified as TIA (International privately owned tank fitted with air brakes). In the lower plate: conforming to the same TOPS code is a large bogie vehicle owned by BRT and used by Shellstar for the conveyance of Anhydrous Ammonia; this vehicle can carry up to 44 tonne and has a tare weight of 35 tonne. *Michael J. Collins/British Railways*

Above: TMV. The only surviving three-axle tank wagons in the private sector, are owned and operated by the Milk Marketing Board (MMB). The fleet consists of 31 three-axle vehicles in the number range MMB 42840-42870 and were manufactured during 1981 by W. H. Davis Ltd of Langwith Junction, utilising withdrawn chassis from ex-BR six-wheeled milk wagons and newly manufactured tanks. A train of these vehicles is seen here at Chard Junction. A similar fleet mounted on four-wheel chassis also operate for the MMB classified as TRV (not illustrated). *B. D. Pibworth*

Below and right: The TOPS code TS is allocated to tank wagons mounted on a two-axle chassis with a GLW of 30-39 tonne, this weight should not be confused with the carrying weight of the vehicles which would be something in the region of 21-26 tonne depending on the manufacture of the vehicle. A vast number of different types used to convey a wide variety of liquid chemicals are classified under this code, and the following illustrations just show a small selection. The first picture is Esso tank No 44358 able to carry 22 tonne of liquid, fitted with vacuum brakes and having a full length tank with central ladder. The second illustration is Esso No 45096, a vehicle able to carry 24.5 tonne, also having a full length tank but with ladder at the end. The last picture is a Lubricant Products tank No 209 carrying TOPS No BPO 4846, this wagon has a shorter than frame length tank with a carrying capacity of 23.3 tonne. *All Colin J. Marsden*

45096
45096
NO NAKED LIGHT

LP. 209
LUBRICANTS PRODUCERS
LIMITED
LP.209

Above: Owners and makers plates affixed to TS group tank wagon. The owner's plate gives their name and address with wagon number below, whilst the adjacent maker's plate tells that this vehicle was constructed by R. Y. Pickering & Co Ltd at Wishaw, Glasgow.
Colin J. Marsden

Below: TSV. It is usual operating practice to form or operate tank wagons in block formations, however some vehicles are offered for use in *Speedlink* wagon load services normally of the modern air braked type. Tank wagon fleets seem to be prone to not carrying their correct TOPS information panels, or even when they do, the code is often missing. The block formation of vacuum braked vehicles here are operated by W. H. Davis of Langwith Junction and carry DAVS private owner prefixes to the numbers. DAVS 48567 is seen here nearest the camera painted in green livery. *Colin J. Marsden*

Above and below: TTB, TTF. Privately owned and operated tank wagons with a GLW of 40-49 tonne are classified TT, with brake fittings of types A, B, F, G, V and W. The two vehicles shown here conform to the standard class B series emerging during the 1960s, having an approximate 8ft 9in diameter tank mounted on a girder channel frame with standard link suspension to the roller bearing wheel sets. In the upper plate is No SUKO 65542, this vehicle is equipped with air brakes and a through vacuum pipe, whilst the lower plate shows No BPO 55384 (also carrying No 3287) being a vacuum braked tank fitted with AFI (Accelerated Freight Inshot). *Both Colin J. Marsden*

Above: TTB. This British Petroleum owned/operated tank No BP 65694 carries a full TOPS information panel under the hazardous chemical label, giving full details of the load carried and instructions for dealing with any mishap that might occur involving its cargo. This vehicle, also of the class B type, is typical of the present oil based product carrying fleet and being in a rather dirty and grimy condition. *Colin J. Marsden*

Below: TTA. This tank vehicle classified under the TT code is not used for conveying oil or petrol, but is owned by BRT and on loan to ICI for use as a caustic soda carrier. The vehicles are fitted with clasp type air brakes and usually operate on the single pipe air system. The maximum load carried is 32.5 tonne and they have a tare weight of 13 tonne. These vehicles are labelled 'Not to be Hump Shunted' due to the nature of their cargo which could cause considerable damage if seepage occurred. *John Tuffs*

Above and below: TTB. Another owner/operator of a number of TT type wagons is Procor, who have a fleet operating throughout the country for various companies, and particularly in the North East. Procor TTB No PR58656 illustrated above is at the ICI Seal Sands chemical complex carrying a cargo of acid. The lower plate also shows a Procor vehicle but of slightly different design labelled for transporting caustic soda. Both these vehicles have mid-way ladders leading to the loading lid and filling equipment. *Both: Colin J. Marsden*

Above: TTA. Storage Transport Systems (STS) are the owners of a number of modern air braked vehicles with a carrying capacity of 32.5 tonne. They are usually painted in blue livery with black underframe, as displayed here on vehicle No STS 53128 photographed when new. *Colin J. Marsden*

Below: TUA, TUB. The final TOPS code for tank wagons is TU, for those weighing more than 50 tonne when loaded. This does not necessarily mean that a heavy cargo is being carried but, as in the case of the vehicle illustrated, the vehicle itself is much heavier than the cargo it can transport. The TUA in the photograph is one of a fleet built and owned by Procor for lease to ICI for the conveyance of hydrocyanic acid of 99.5% purity. Their latest designed air braked vehicles usually operate in small block formations between large barrier wagons. Both ends and sides of these vehicles carry notices advising people who are not directly involved with their movement and use to keep well clear, together with telephone numbers from where assistance may be obtained should any lethal cargo escape. PR70175 was photographed at Seal Sands. *Colin J. Marsden*

Freight trains

The following eight illustrations have been included to show some of the types of wagon included in this book in the usual context that the reader will observe them, ie: behind a locomotive operating on the main line. Some vehicles will have minor modifications which are not shown in this book but the basic type of vehicle and the cargo carried should be identifiable.

Above: The HAA or modern air braked hopper wagon for transporting coal will usually be seen operating in block train formations. The train illustrated is headed by two Class 26 locomotives and consists of HAA vehicles with canopy tops, rarely seen outside Scotland, although when the vehicles were first introduced a vast number were so fitted. *Ian Bald*

Below: This view of an unfitted freight train hauled by a Class 47 locomotive, consists of two air braked ferry vans, a vacuum braked STV *Tube,* three covered grain hoppers CGV, one wooden sided *High* of the OWV group and a brake vehicle, CAP type.
Colin J. Marsden

Above: Running as a block load train with a brake van at either end, is a set of CCP *Covhop* sand hoppers being used by British Industrial Sand for Rockware Glass. The leading brakevan is of the CAR type identifiable by the yellow panels on the side and ends. The train is passing Barnby Dun, near Doncaster, headed by a Class 31 No 31.174.
Colin J. Marsden

Below: Nuclear flask train. It is usual practice to form special freight trains when flasks require transporting, thus avoiding the conveyance of flasks in regular freight services. In this view a Class 25 hauls an XKB type flask marshalled between two barrier vehicles (formerly SAAs) with a dual brake fitted brake van on the rear. The train was photographed at Llandudno Junction while en route from Trawsfynydd to Windscale. *John Vaughan*

Above: The uniform block load train is an ever increasing sight on the railway network today. Photographed between Maidenhead and Twyford a Class 47 hauls a block train of PGA hopper wagons owned by the Tilbury Group towards the west, returning to the Westbury area where the stone quarries are situated. Normally trains of PGA type vehicles operate in 20-25 wagon formations. *Colin J. Marsden*

Below: Motor cars booked to travel by the BR Motorail service are transported in two ways, either on a flat open vehicle classified under TOPS as FVV/FVX depending on the brake fitting, or in a covered wagon of type NX (dealt with in *Rolling Stock Recognition: No 1 Coaching Stock*). The regular service between Paddington and St Austell is formed of FV type vehicles and is seen here headed by a Class 47 near Hele and Bradninch. When motor cars are loaded on these open flats windscreens are supposed to be covered with rubber panels to prevent damage, but this is not always carried out. *Colin J. Marsden*

Above: Today in the age of air braked block and *Speedlink* services one seldom sees the good old fashioned unfitted freight train formed of a mixture of vehicle types. This long unfitted freight headed by Class 40 No 40.022 is seen coming off the Settle & Carlisle line at Settle Junction on 7 February 1983 with the 13.10 Carlisle-Healey Mills freight, formed of type MCV, VEV, HTV and PCA vehicles. *Colin J. Marsden*

Below: For a number of years a regular sight trundling through Devon and Cornwall has been the china clay trains formed of 12.5 tonne clay vehicles of the UCV type. In this view at Dawlish Warren Class 50 No 50.039 *Implacable* hauls some UCV vehicles towards the west. The train also has MCV, VVV and MDV types in the formation. *Colin J. Marsden*